The Janos People

GLORIA   FRANCES
JOHN
TANYA   NATASHA

FRANK JOHNSON

The Janos People

*a close encounter
of the fourth kind*

SUFFOLK
NEVILLE SPEARMAN

First published in Great Britain in 1980 by
Neville Spearman Limited
The Priory Gate, Friars Street, Sudbury, Suffolk

Copyright © Frank Johnson 1980

ISBN 0 85435 374 7

Photoset in 11/12 pt Plantin by
Galleon Photosetting, Ipswich
Printed and bound by
Biddles Limited
Guildford and King's Lynn

Contents

page

Author's preface		
Prologue	Disaster in another world	1
Chapter One	Encounter in England	8
Two	John and Frances dream	21
Three	Natasha's testimony	31
Four	Entry into the spaceship	46
Five	Frances examined	53
Six	What Frances was told	63
Seven	John examined	79
Eight	John in the engine room	96
Nine	Pictures in the navigating screen	107
Ten	Underground encounter	120
Eleven	Departure from the spaceship	126
Twelve	The Janos people	136
	the planet Janos	137
	climate and vegetation	139
	buildings and transport	140
	power and industry	145
	food and animals	148
	clothing and hair styles	148
	flags, badges and insignia	151
	speech and language	154
	telepathic communication	155
	personality and politics	156
	physical type and race	156
Thirteen	Under whose flag?	158
Fourteen	Homecoming	176
	An open letter to the *Janos* people	189
	A *note on credence and credibility*	191
	Books for further reading	193
	Where to report incidents	194
	Index	195

List of Illustrations

page

Frontispiece: The Family

1	Site maps: lift-up and set-down	10
2	Natasha's drawing of Phusantheas	44
3	Room layout and chair detail	61
4	Pleasure boats and pennant detail	72
5	Domestic dress and hair styles	74
6	John's medical room	81
7	John with Serkilias	83
8	Oscilloscope patterns	85
9	Badge worn by spaceship crew	89
10	View across the engine room	94
11	Section through engine room	95
12	Plan of engine room	98
13	Transformer? unit	102
14	Profile of a rotor cylinder	105
15	A flight of asteroids	112
16	Float-vehicle	117
17	Section through tunnel	119
18	A private home on Janos	142
19	A nuclear power station	144
20	The Janos 'flagship'	146
21	Janos woman's evening gown	152

All the illustrations, including the photograph frontispiece, are by the author, assisted in many cases by sketches and descriptions provided by Frances and John.

The drawing of Phusantheas is an enlarged copy of a detail from a sketch by Natasha.

Author's Preface

IN PRESENTING THIS account of the Janos people, a human community which has been for many thousands of years away from the Earth, living on a remote planet, I am very conscious of presenting something which will upset many comfortably-established habits of thought.

The reader may accept my account as factual, or reject it, as he may choose; it does not concern me one way or the other. The story only seems improbable because it throws a hefty spanner into the works of accepted assumptions about the history of mankind: technically it hangs together in a remarkably convincing way.

It will also upset many ufologists – people who study and enquire into the strange phenomenon of the UFO or 'flying saucer'. Ufology, no less than other, more respectable fields of specialised knowledge, has generated a lot of theories, some of them, to my mind, highly unlikely, and, indeed, in some cases strictly meaningless.

This account, the first published presentation of a straightforward, realistic human background to the flying saucers, sweeps away all the armchair theories, if you take it at its face value; and I am confident that anyone who reads this book without preconception and with an open mind will end by taking the account at its face value.

However improbable the Janos story may seem at first acquaintance – and it is sensational in a way that fiction could not rival – one must consider what probability attaches to any alternative hypothesis. The whole question of probability and credibility will be examined at the very end of the book, by which time the reader will be in possession of the facts, and better able to judge for himself (the male, of course, as always, embracing the female; the English language lacks a common gender, such as Greek possesses, for these general references to humanity, irrespective of sex).

This is the story, as told in my presence, of an English

family of three adults and two children, who spent nearly an hour as the guests of the Janos people, on board one of their spaceships, over rural Oxfordshire. One of the senior officers told them: *"Our ship is one of those which have been chosen to make the first contacts"*. (Throughout the book, *italic* type is used to denote direct verbatim quotations from any of the Janos people in the spaceship; it implies that these were the actual words used, and not a paraphrase.)

The Janos people are, in the literal sense of the term, extra-terrestrial: they come from another world, which they call Janos, a planet *"several thousand light years"* away from Earth. At the same time, they are terrestrial in origin: their remote ancestors, in the prehistoric past, lived here and originated here; the people seen in this particular incident appeared to be Europeans of a Nordic type.

This is surprising; but it does not in itself violate probability. It does raise some rather difficult questions, as to how? and why?, which will be discussed towards the end of the book.

This investigation was undertaken at the suggestion of Jenny Randles, Secretary of UFOIN (UFO Investigators' Network), an organisation of limited membership and high standards. I must make it clear, however, that neither Jenny Randles nor UFOIN has any responsibility for the investigation, or for the conclusions I have reached. It has been, throughout, a solo job; my only colleagues have been the five members of the family whose unusual story is told here, and our hard-working hypnotist.

As explained in the text, hypnotic regression, or regressive hypnosis as it is variously called, was employed as a method to aid the investigation, for a dual purpose: to dissolve away the amnesia which (in common with many close encounters) has been a feature of this case; and to enable the witnesses to re-experience episodes within the encounter, several times if necessary, to obtain a more detailed and more sharply defined recall than normal memory alone could give.

The success of the investigation owes a great deal to the professional skill and untiring patience of Geoffrey M'Cartney, hypnotherapist and consultant hypnotist in the city of Gloucester. At the same time, it must be stated that the case

Author's Preface

does not depend on hypnosis alone; very many items of information, including some of the most important, were obtained in normal recall on occasions when the hypnotist was not present.

I must also pay tribute to the members of the family, especially John, Frances and young Natasha, for their willing cooperation. They have been very strongly motivated to try to understand what happened to them; nevertheless their readiness to submit to seemingly endless questioning, both under hypnosis and otherwise, has been a major factor in bringing to light the story of the Janos people. The fact that their recorded interviews amounted to forty-seven c90 cassettes is an impressive statistic on its own.

One thing which has distinguished this investigation has been its sheer size and complexity. Many 'encounters' can be fully described in half a dozen pages; whereas this experience, of less than an hour, has yielded so much detail that I have had to condense drastically to bring it within the compass of a book.

This enormous yield of detailed information has been the reward of sheer persistence on all our parts: it would have been so easy, at many points of the investigation, to be satisfied with what we had, and so miss most of the story.

During the thirteen months of the active phase of the investigation, I have come to know the family very well, and have acquired a considerable respect for the integrity and insight, especially of the two principal witnesses John and Frances, who are brother and sister, with whom I have spent a great deal of time.

By the wish of the family, I have withheld their surnames and postal addresses. This is in accordance with the spirit of the advice given them by officers of the Janos spaceship, who imposed a degree of post-hypnotic amnesia on them, to protect them from the undesirable consequences of immediate publicity. (*"It will cause you much trouble."*)

The witnesses have checked their own statements, and have occasionally made a minor correction. The conclusions, however, are my own; and they hold no responsibility for them, though they are in general agreement.

I have been greatly helped by the many sketches made by

John and Frances, which I have used as a basis for the illustrations. John is reliable and accurate on the shape and arrangement of objects, and has a good sense of layout and orientation; Frances is remarkably good at remembering speech, word for word, and has an excellent memory for detail; while Natasha, for her age, is an intelligent and observant child, whose contribution to the investigation has been by no means negligible.

It has been a great pity that Gloria's amnesia has been so severe and refractory. She is beginning to have odd flashes of visual recall; doubtless her memory of the incident will return fully in time, and will provide a valuable cross-check on the testimony of the others. Tanya has, without doubt, a full memory; but as she was only three years old at the time, she cannot be regarded as a formal witness.

I do wish to make it clear that I make no claims that this book solves all the mysteries of ufology. It gives a detailed, but not a comprehensive account of the Janos people, limited to what one family have been told about them, and to what I think can be legitimately deduced from that telling; but it does not go outside this theme. I think we shall find that many reported encounters, when re-examined in the light of what this book has to report, will be found to be part of the Janos story, especially where they deal with fair, blue-eyed normal people in silver uniforms; but it will be apparent to anyone familiar with the literature of ufology, that there are many other reported incidents which do not seem to fit, as far as our present knowledge goes.

Perhaps the Janos people are not the only visitors to our planet from other worlds. Perhaps the Janos people themselves, when we are able to ask them, may be in a position to throw light on these other extra-terrestrials.

I will now take you back in time some thousands of years, and away in space a few thousand light-years, to a remote planet called Janos by its people, where disaster hangs in the sky.

Frank Johnson
Worcestershire, England
April, 1980

PROLOGUE

Disaster in Another World

JANOS* IS THE name of a world that died.

Janos was a planet, very like our own, with blue seas and lakes, green fields and hills, trees and grass; with towns, cities and quiet countryside; with ships and boats and aircraft; with pleasant single-storey homes and families of men, women and children, people very like us. People who laugh and make jokes; gentle, kindly, sensitive people of deep understanding, who abhor violence and will not make war, even to obtain that which they most desire. People who, even in extreme need, do not want to bring trouble on others. People very like us; but perhaps in all honesty we ought to say, the kind of people we would like to be, rather than the kind of people we of this planet all too often are.

Janos had two moons, both small compared with our moon. One of these, the nearer one, called Saton†, imperceptibly slowed, millenium after millenium, by the friction of solid tides raised in its rocky crust by the gravitational field of the planet, had crept nearer and nearer to Janos, until it was too close for stability; the cohesive forces which held it together were finely balanced against tidal disruption.

The people of Janos had known for a long time of their danger. For a long time they had known that they must eventually leave the planet, as the crew and passengers of a sinking ship must take to the boats.

With a science and technology more advanced than ours, they long ago developed the art of space travel, and constructed spaceships in which they explored the worlds within

---

* pronounced 'Jáne-oss', the first syllable stressed
† pronounced 'Záton'; but the spelling Saton was given by the Janos people

reach of them, looking for a new home, to which they could go when the time came to leave Janos.

Their exploration ships ranged far and wide among the nearer stars and their attendant planetary worlds. This interstellar househunting took a long time; and the crisis crept nearer. Many planets were surveyed, and some of them would have made possible homes.

But one green, lovely world had attracted them from the first, although they knew it only by tradition; because, they tell us, it is so like Janos as it used to be. There was a feeling of home about it. They had known about Earth for a long time, from their history books; they knew where it was in the sky.

There are differences: both Earth and Janos have land and sea; but whereas Earth's oceans are far more extensive than its continents, in Janos the land areas are greater than the areas covered by seas and lakes. The climate of Janos was mild and gentle; those who have seen the planet from space saw nothing to suggest snow or polar ice.

The few scenes we have of the old Janos paint a picture of sunlit greenery and blue water, flanked by low wooded hills. On the lakes, gaily coloured power boats, pennants fluttering, throw up their smooth bow waves, ripples fanning out towards the shore, where fashionably dressed women, with their menfolk, enjoy their abundant leisure. Nearby, a peaceful community of modest, pleasant single-storey homes have flaxen-haired children in them, and white-fenced gardens around them – a scene so familiar that it touches the heart. And all this was thousands of light-years away.

The Janos people have known about Earth for a very long time – long before a double disaster drove them from their home world. For besides the threat of the fragmenting moon, which they had long warning of, an unsuspected danger lay in wait for them, for which they were totally unprepared; this other peril we will come to presently.

Like everyone else with long warning of the need for action, they ran it a little too fine; and the rain of great rocks began before they quite expected it. They were almost ready: out in orbit, above the fragmenting moon, the great fleet lay in its gravitational anchorage, a host of mighty ships, clustered about one really stupendous ring-shaped flagship, a veritable city in space, capable between them of taking in the entire population of Janos, many millions in number, and provisioned and ready for a long interstellar voyage. They had already picked out their destination.

The big ships, built in orbit, could not embark directly from the ground; and hundreds of smaller vessels were needed to ferry the population up from planet to fleet, making endless to-and-fro trips, each time carrying a full load of men, women and children, with all their stores and necessities. These smaller vessels, lens-shaped ships of the kind familiar to us as 'flying saucers', were built in huge subterranean shipyards, far below the threatened planetary surface, in vast artificial caverns whose lofty roofs were supported by massive columns hewn out of the solid rock, columns which flared out trumpet-wise at the top, to meet the curving roof.

The operation of embarking a whole world of people was well advanced when, ahead of estimated time, a series of minor rockfalls gave warning that the inner moon, Saton, was about to break up. The English family whose story is told in this book have been shown films taken during one of these earlier rockfalls: they describe rocks "as big as houses" falling from the sky, crushing the buildings and setting them on fire. The film's sound track records the screams of people trapped in the blazing wreckage.

A curiously unexpected detail was caught by the camera. Seated upon the edge of a rocky cliff, gazing up into the sky, a big dog – a wolf, rather – howls inconsolably at the falling rocks. It is dark-coated, not quite black, with long, thick shaggy hair which rises in a fear-generated crest over the head and spine. It has the look of a wild animal; what it is doing there in Janos is a bit of a mystery, which we will go into later: events just now are moving too quickly; we have no time to speculate.

The early beginning of rockfall created a complicated, fast-moving crisis. Those who were not killed outright made, if they could, for the nearest tunnel-mouth; these were entrances, of two very different sizes, to the system of tunnels which led down, far down into the planet's crust, to reach the complex of subterranean shipyards. The small tunnels were intended for workmen to come and go.

Underground, the ships which were in the shipyards and ready to fly hastily took on board as many people as they could cram in, and flew out through the great tunnels, shaped to allow a lens-shaped ship to fly through them with a little room to spare, up to the distant surface, risking the rain of rocks, any one of which would have crushed it like an egg hit by a cricket-ball, and up into orbit with the fleet. Ships returning to the ground from orbit repeatedly entered the tunnels to take on more refugees, until the battering of the rocks caused the tunnel-mouths to cave in dangerously, and the rescue operation had to be suspended until the end of rockfall, several months ahead.

We do not know how many were left underground: they should have been safe enough during the main rockfall which now began as Saton slowly disintegrated, covering the planetary surface with millions of rocks, great and small; for deep underground, the rocks could not hurt them, and they had supplies awaiting transhipment, and uncompleted ferry ships to live in.

But there was a complication which the Janos people now admit they had completely overlooked. The world of Janos was one which used a lot of electrical power to supply its highly-automated industries and its domestic needs, more developed than ours. The main, if not the only source of energy was a network of huge nuclear power stations using uranium, scattered over the planet on the exposed surface; wisdom after the event would no doubt have put them deep underground. It had been assumed that they would be destroyed by the rockfall, like everything else, and would simply cease to operate. With the destruction of all surface industry and the abandonment of the planetary surface, electricity would not be needed: almost certainly the subterranean shipyards had independent emergency generators of low power, which would provide some lighting and ventilation for the entombed survivors.

But it was not as simple as that. The ship people showed one member of our English family pictures of what happened. She saw, first, a typical Janos power station, so sharply defined that she was later able to make a detailed drawing of it; she described it as "like a gasometer inside the Eiffel

Tower" – a very apt description. It is shown in figure 19 on page 144; you will see a massive dull grey cylinder supported within a four-legged tapering pylon of shiny metal lattice-work construction.

Then, as she watched, the whole structure disintegrated in a tremendously brilliant flash. A second or two later, there was another flash, far away in the distance; then another still further away; and more and more flashes, very remote. The Janos man who showed her the picture said: *"There was a chain reaction"*.

All the power stations of Janos exploded, one after another very quickly; as one disintegrated, we are told, it caused the nearest also to explode, and that detonated the next, and so on. What exactly went wrong in terms of nuclear engineering, we cannot explain in detail; but it seems likely that the massive protective shell, which had withstood the earlier, lighter rockfalls, was finally ruptured by the later tremendous battering, and that the control mechanism was wrecked, causing the pile to run out of control and to 'go critical'. Engineers must explain this more fully than I can.

What is quite clear is that the heavy biological shielding was explosively ruptured; and that thousands of tons of intensely radioactive dust swirled around the planet with the turbulent winds created by the multiple explosions. The swirling dust clouds were seen in later pictures; and we know that a lot of this heavy, gritty dust, sucked down no doubt by ventilation machinery still operating, found its way down the tunnels into the underground shipyards, where the people left behind were still living. The entire planet, surface and underground alike, was drenched in radioactivity.

Communications between the planet and the fleet in orbit had failed with the loss of power; and the people in the ships had no idea what was happening in the shipyards. They were in orbit, waiting, for a long time; when the rockfall finally ceased, they still thought it would be possible to send rescue ships down to pick up survivors, and they sent powerful anti-gravity equipment capable of lifting heavy rocks and clearing the choked tunnel-mouths.

When the rescue ships finally penetrated to the under-ground shipyards, their crews were horrified to find that, although many were still alive, they were already doomed to a slow death by severe radiation poisoning. Not one could be saved. They could not be taken in by the rescue ships; it would have been dangerous, even to touch them. They had to be left to die.

The rescue ship crews did what they could to organise for the needs of the sick people, without becoming themselves contaminated. Somebody invented a new type of clothing – rather like a monk's long-skirted habit, with a deep hood or cowl over the head – to give better protection against the lethal dust. The sad routine began, of regular daily visits to collect the dead bodies, and remove them for disposal.

As the survivors became fewer in number, they were concentrated into a few localities, until finally only one shipyard-refuge still had men and women living in it. At this point, a film record was made, by a camera mounted behind the windscreen of one of the last float-vehicles to make its melancholy rounds; the film shows a half-dozen of those still able to walk, carrying a roughly-made coffin to the vehicle, where it was loaded into a cargo-space underneath. While it waited, the vehicle still hovered a few feet above the ground.

The bearers looked like very old people; they shuffled slowly and hopelessly about, "as if they had given up". Some were blind; all had lost their teeth, so that their cheeks were sunken; and their hands were deformed and claw-like, with huge lumps on the knuckles.

The ships remained in orbit until the last of the sick people died. Then the great migration fleet moved slowly out of its orbital station, leaving Janos for the last time. Gradually its speed built up, as the navigators set course for Earth's star – what we call the Sun, thousands of light-years away. They knew its galactic coordinates, and how to find it. The acceleration was maintained until the fleet was flashing across the Galaxy at a velocity close to the speed of light. In this way, a journey which, as measured by Earth time, took many thousands of years was completed, according to the ships' clocks and calendars, in just two years.

The traumatic stress of the double disaster – especially their grief over those who were left to die – affected the Janos people deeply; but gradually, as the voyage went on, their natural buoyancy of spirit returned. Today, they show the emotional effect of all that befell them – loss of their home, the rock fall, and their shocked grief over the fate of those left behind – only when talking of these things to visitors, and showing them films, brings it all back.

CHAPTER ONE

Encounter in England

JOHN, HIS WIFE Gloria, their daughters Natasha (then aged five) and Tanya (then aged three), and John's sister Frances were travelling together by car from Reading, where they had been attending a family funeral, to their home near Gloucester. It was late evening, Monday 19th June 1978. They followed a route familiar to them, 417 through Wantage, Stanford in the Vale, Faringdon and Cirencester.

After Stanford in the Vale, they began to notice a bright light in the sky, in front of and somewhat to the right of them, which seemed to keep station with the moving car. John was driving at less than fifty miles an hour, and it seemed impossibly slow for an aircraft. They discussed it idly: could it be a helicopter? It did not seem likely; and they could not hear anything.

John suggested, more as a joke than anything, that it could be a flying saucer. John is rather given to joking, and nobody took much notice of this. They drove on, still watching the light. Presently they noticed a smaller red light to the left of the white light; their constant relation suggested that the two lights were carried by a craft of some kind, which they could not see, though the sky was clear, and there was a full moon high in the sky. They were still thinking it must be some kind of aircraft, but were puzzled by the way it seemed to pace the car.

John said he was going to stop and have a look at it, to see if he could make out what it was and what it was doing. He slowed down; but Gloria said: "You can't stop here; there's a house by the side of the road all lit up, and the people might think we were funny stopping outside, and come out to see what we are doing". John accordingly drove on a little way;

and then he determined to stop, come what may. He said he was going to stop, even if a police car was coming. This was an odd thing to say, because there was no reason why he should not stop by the roadside if he wished to. The road was quiet, with only occasional traffic.

Gloria's reaction to the lighted house was also a little illogical; because if there were anything untoward, they might feel that the presence of other people would be reassuring. But it seemed logical at the time.

There is something odd about the lighted house; because the next time they travelled along the same stretch of road, a few days later in daylight, they were able to pinpoint the exact position where they had seen the house – but there was no trace of a house anywhere near, although as they remembered seeing it, it was only some fifteen to twenty feet back from the road.

There was nothing remarkable or unusual about the house they saw that night, except that it simple wasn't there. It was just an ordinary house, of a type not uncommon in the area; in fact, much later on, on another journey along the same road, they realised that they had already passed a house like it – but this one was on the other side of the road, and some way further back; that is to say it was further east, and on the right hand side of the road as they went east to west; whereas the 'house all lit up' that Gloria spoke of was on the left hand side of the road, going in the same direction.

Later, under investigation, John and Gloria independently made drawings of the lighted house, which agree closely. They show two windows upstairs and two downstairs, all four windows, they said, being uniformly lighted as if a translucent screen were drawn across them. No curtain or interior details were visible. On the occasion much later, when they first realised that they had already passed a similar house on the other side of the road, one of the four windows was lighted, and it was filled by a translucent blind which completely filled the opening, showing no detail.

Later in this book, in Chapter 11, the reader will find that there was another occasion when visual impressions were transferred from one side of the road to the other. I feel that, in the present incident, what probably happened is that the

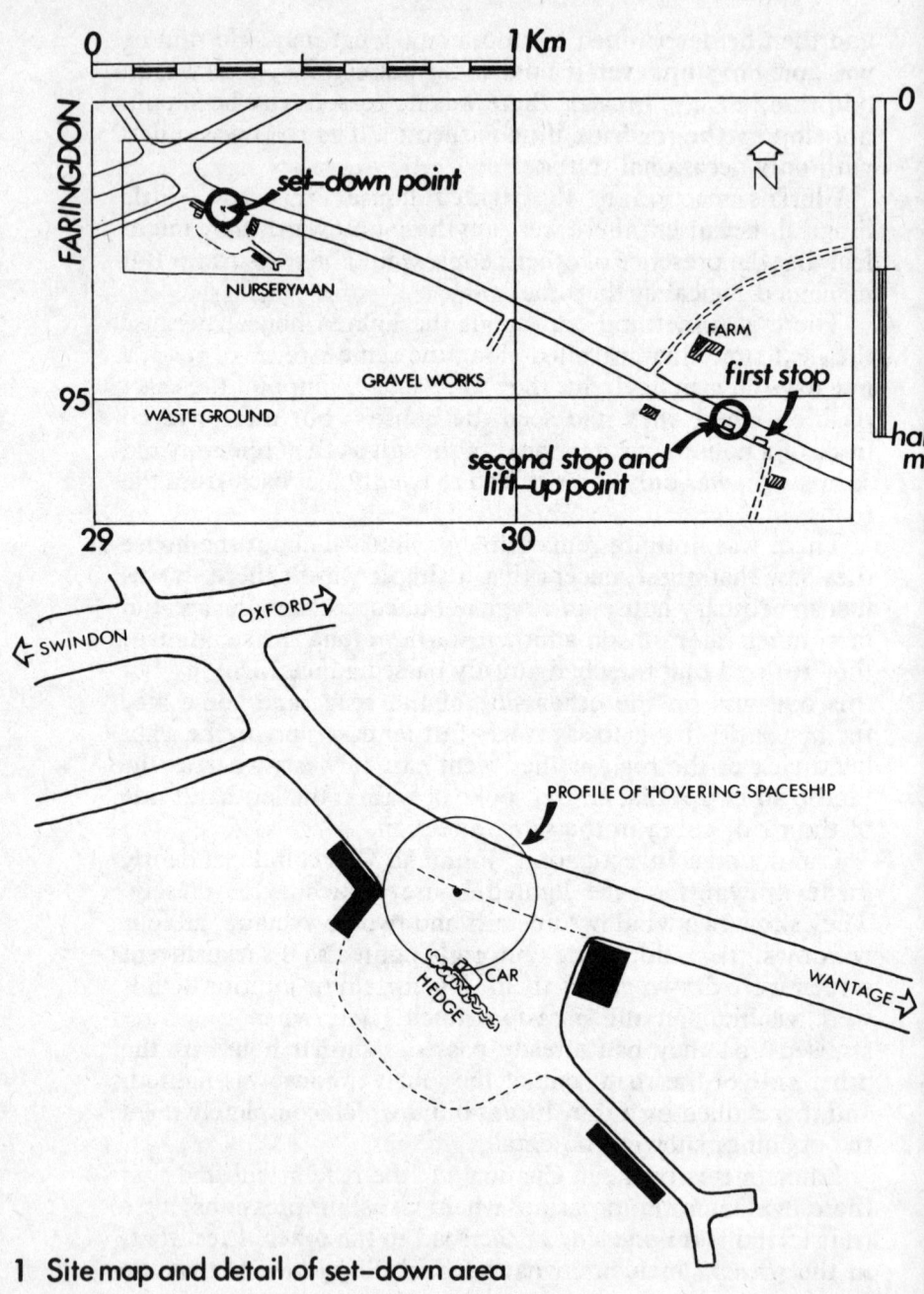

1 Site map and detail of set-down area

people in the spaceship, who were already watching the car and probing into the minds of its occupants, gave them a hypnotic suggestion that they should see again the house with four lighted windows (which they had already passed without at that time consciously noticing it), but that they should see it on their left – the side the car would draw up to – and at the same time feel in their minds that this was not a good place to stop. I think the flying saucer people, for their own convenience, wanted the car to stop at a point which they had already decided upon, a little further down the road; and when they came to it, John was given a strong urge to stop the car.

To continue with the story: the light in the sky seemed now to be approaching them; John pulled up by the roadside, and, leaving the engine running and the headlights on 'dip', he got out of the car and looked in the direction where the light had been seen, but at first failed to find it. (Non-British readers should remember that in the United Kingdom, traffic drives on the left; this is necessary to an understanding of this part of the incident. John was thus standing on the right of the car.)

John scanned rapidly up and around to find the lights, and in a moment saw that the craft had moved very quickly into a position directly in front of the car, but up in the air, at an elevation of about sixty degrees of arc. He could now see the shape of the craft; it appeared, he said, to be a very large circular object, completely black, the under surface, all that he could see, being curved like a shallow bowl seen from below. The white and red lights could not now be seen. There was no visible detail of any kind on the craft. The bowl shape was more curved in its central part, with a flatter extension all round it.

The circular craft now moved quickly over the car, and away to the right of the road, somewhat behind the standing car; it sank down to a low level behind a thin row of trees, but could still be clearly seen through the gaps between the trees. A row of small coloured lights now appeared all round the extreme rim of the disc; the lights were of different colours, the same colour not appearing twice in succession. The whole ring of lights was revolving slowly from left to right. The craft rose and fell slowly several times, but did not quite reach the ground. Gloria and Frances could both see the craft through the rear right hand side window of the car. In profile it was a typical flying saucer – a biconvex lenticular disc, tapering at the edges to a fairly sharp rim where the lights were; the central part was more curved, both above and below. It was all dull black except for the lights.

It is of some interest to note that more than a year later, in mid-August 1979, a local newspaper stated that a taxi-driver had reported to the police that he had seen a UFO at about 2 a.m. over Stroud in Gloucestershire, quite near the home of one of our witnesses. He described it as a large disc surrounded with multi-coloured lights, moving slowly and silently. Even this sketchy description is enough to suggest that this was either a re-visit from the same spaceship, or one very like it. A similar spaceship was also reported, on a subsequent occasion, over Cirencester, not far away.

Returning to our story: during this time, a sound was heard, coming from the craft, which the witnesses described as a mixture of two different sounds: one was a soft, sibilant rhythmical sound, like someone whispering "swish – swish – swish"; the other was a continuous metallic drumming noise, like that of a distant railway train. These sounds were quite distinct from the familiar quiet purr of the idling car engine. It is an unusual feature of this case, that at no time did the close proximity of the flying saucer cause any malfunction of the car's electrical systems; engine and lights functioned normally throughout.

The flying saucer's downward movements, close to the ground, alarmed Gloria, who said: "John, get back in the car and let's go; it's going to land". Throughout the whole encounter, Gloria was frightened, though by no means hysterical; of the five, she was the only one who experienced any fear.

John did so; and drove on. By now, the children were awake; Natasha said: "Is it a naughty boy with a kite?" She said this just as the car began to move.

From this point, the recollection of events by the adult witnesses divides into two parallel and different experiences, which reunited when they reached Faringdon. We will distinguish them as the 'real story' and the 'cover story'. At the time, only the 'cover story' was remembered; but even at the time, it had a feeling of unreality, and the witnesses found it strange, puzzling and unconvincing as a real experience. We later came to the conclusion that it was a synthetic memory, put into their minds by the flying saucer people to account for the loss of time; the point of this will emerge presently.

In the 'cover story', as soon as John drove on, they were all puzzled by a total change in the character of the road. One should remember that the road was thoroughly familiar to them, from the regular visits which these five people made to the parental home near Reading. It could be said that they knew every inch of the road. The deception, whatever its purpose, might have succeeded with strangers who did not know the road well; but to this family, the departure from reality was sharp and unmistakable.

The road they now found themselves driving along was very much narrower than the road they knew, and was closely set about with tall hedges, too high to see over; whereas the sides of the real road are open, giving extensive views across country. All three adults were aware of a sense of strangeness and unreality.

John was very uneasy, and could not understand how they came to be on an unfamiliar road, since there was no possibility of his having taken a wrong turning; the real road went straight to Faringdon, little more than a mile away, without any branch or junction.

But this narrow, closely-hedged road seemed interminable. It seemed to all of them to go on and on: the two women had a floating sensation (characteristic of some hypnotic states); and all felt the car's movements to be unnaturally smooth, as if it, too, were floating along, not in contact with the road, which looked rough and ill-made. The real road is broad, smooth, well-maintained, and almost completely level and straight; but this narrow dream-road (as it seemed) went up and down hills, and curved to left and right.

There seemed to be a lot of repetition in the details of the road; there was a characteristic pattern of a bend at a rise, followed by a dip, that repeated endlessly. The hedges and trees which lined the two sides of the road seemed the same, as if they were mirror-images.

John had a sense of having gone back in time. He said it was as if someone had been told: "Paint a picture of a country lane"; it was old-fashioned and conventionalised – note the similarity, in this respect, to the 'house that wasn't there'.

The car belonged, not to John but to Frances; but John was driving because Frances would have a further journey to make that night. He was not familiar with the car, and had driven it for the first time that week-end; this journey was the second time he had driven it. Frances said afterwards that the car seemed strange and unfamiliar; "it didn't seem to be our car at all".

John seemed to have no control over the car; he said afterwards that he felt that if he had taken his hands off the wheel, and his feet off the pedals, the car would have driven itself. When the car climbed a hill, he did not change gear or press the accelerator down for more power; and when it swung left and right, although he was aware of turning the steering wheel, it seemed not to need his volition.

During this curiously unreal and long-drawn-out journey, they could see the single white light of the UFO following them; Frances at the time found this reassuring; she said later: "I didn't feel that I needed safety; it wasn't that we were being chased by something that was going to hurt me – just that we were being followed". This makes sense only if one assumes that this unreal drive was planted in their minds after the whole of the spaceship experience was over, but appears at this stage in their memory sequence to account for the time lost. A remark made by the ship's captain in his speech of welcome (reported at the end of Chapter 4) has some bearing on this.

Subjectively, they all agreed when questioned later, this curious 'journey to Faringdon' had seemed to take well over half an hour, perhaps three quarters of an hour. No one had thought of looking at a watch – this would have given it away as an unreality – and, although John repeatedly glanced at the speed indicator "to see if I was making any progress", it did not occur to him to look at the mileage recorder. The speed indicated did not vary; it stayed at a figure between 35 and 40 miles an hour.

Suddenly, without warning, they were in Faringdon; it is

significant that they failed to see the very conspicuous sign bearing the town's name, a hundred yards before entering the town. Their minds were so much on Faringdon, that as the name flashed up brightly in the car's headlights, they could not have failed to notice it; but we learn from the rest of the story that they were never on that section of the real road at all, and did not pass this sign; their real re-entry into the road was beyond the sign.

Everybody greeted the lights of the town with cries of relief. "Civilisation at last," said John; "I thought we were never going to get here". Faringdon, a small town they knew well, seemed unusually quiet, with hardly anybody about; they did not realise that the time was nearly an hour later than they imagined it to be.

They passed through Faringdon; then, as they came out into open country, Gloria, who was watching through the rear window, reported that the light was still following them. They all looked at it; John took a quick glance over his shoulder and saw it. Frances said afterwards that each time they passed through a village or town, the light disappeared, then reappeared again when they were once more in open country. She is positive that it was not a case of the light passing behind buildings; it was switched off as they approached any inhabited place, and switched on again after they had left it.

The light followed them all the way to Cirencester, about eighteen miles beyond Faringdon; after Cirencester they lost it, but it did not lose them, as we shall see presently.

They arrived home (that is to say, at John and Gloria's home) without further incident. It was not until they were inside the house that anyone realised that it was very much later than they thought. The journey was so familiar to them that they always knew, reckoning from the time of their departure from Reading, at what time they would arrive, generally within a few minutes. On this occasion, their 'expected time of arrival' was about twenty minutes past eleven. Instead, the house clocks showed fifteen minutes past midnight. John used the telephone to check the time. In fact, the 'interminable' imaginary journey through the narrow, closely-hedged lane would have lengthened their journey by something like this amount of time, had it been a real experience; but it is clear that the family were, at this stage, tending not to accept it as a real experience, and they were accordingly puzzled by the loss of nearly an hour.

It should be explained that none of the family had any previous knowledge of flying saucers, and had read no books about them, though John and Gloria (but not Frances) had seen the film *Close Encounters of the Third Kind*. Frances, particularly, later confessed that at the time of their own encounter, she did not believe in flying saucers and had no interest whatever in such things. None of them were aware that in published accounts of CE4s (close encounters of the fourth kind, such as the one they themselves had experienced, though they had not yet recovered memory of it) it is often the loss of time unaccounted for which first alerts investigators to the possibility that a CE4 has taken place; loss of memory, partial or total, is a frequent, but not invariable feature of such cases, and we shall learn presently how and why this occurred in their own encounter.

John then telephoned a Royal Air Force station in the area of the encounter to ask if they could throw any light on their sighting. They were still reluctant to believe it was a flying saucer, and thought they might have seen some experimental aircraft with a circular appearance. The RAF took down the particulars, and asked some questions. Later they told John that they had made extensive checks with other service airfields, with the civilian airports, and the police; they were satisfied that no aircraft of any kind had been in the Faringdon area at the time. However, they did say that several people had telephoned in to them and to the police, to complain of a low-flying aircraft in that area.

Frances finally went on to her own home near Stroud, about nine miles away. It was now very late, and John and Gloria tried to persuade her to stay the night with them; but she was anxious to return to her own family. Before she left, Natasha said to her: "Auntie Frances, be careful you keep your windows tight shut; or you might get sucked up into a spaceship".

John and Gloria went to bed, feeling rather sick, as if something had disagreed with them. They were still awake when, about half past one, they heard the familiar and unmistakable sound of the flying saucer (the same flying saucer, they insist) passing slowly over the roof of the house.

During the next few days, all three adults (but not the children) experienced itching of the skin: John and Gloria itched on the arms and legs; but Frances did not itch until the Friday of that week, when she went to the hairdresser, and the shampoo stung her scalp painfully.

Another physical symptom developed at this time which affected all three adults, but again, not the children: all three developed marks similar to bruises in appearance but more sharply defined; the discoloured areas did not hurt when pressed, as a real bruise would. Frances had a 'bruise' on the outer side of the lower right leg, oval in shape, about three inches long, dark blue in colour, which remained for two days, then "just went". Gloria also had a dark blue mark, smaller than the one on Frances, and round in shape, situated just below the right knee on the inner side; this also disappeared after two or three days. John had a mark similar in size and shape to Gloria's, but on the outer side, just below the right knee; this was not blue but light brown in colour, and was one inch in diameter.

Two months later, I mentioned these marks to a physician with 'close encounter' experience; he suggested that, in association with itching, they may have indicated prolonged contact with some object which was slightly radioactive, and advised that blood tests be carried out as a precaution. This was done by arrangement with the family doctors concerned; both red and white cell counts were by that time found to be normal.

As we shall learn later, John did have something clamped around his right leg, just below the knee, during the medical examination which commonly features in such experiences; Frances remembers quite a lot about her own 'medical', but she does not remember a leg clamp.

We have no positive certainty about what happened to Gloria, though the fact that she had similar symptoms to the other adults suggests that the experience which produced the symptoms was also similar; but Gloria's amnesia has been so strong and so persistent that she herself remembers very little, and (since the other adults were separated) we depend mainly upon Natasha for an account of her mother's examination. She only saw the beginning of it, for she was also taken to another room; but she stayed long enough to see her mother laid on a couch, and that a clamp was attached to each of her legs, just below the knee, with wires connecting the clamps to instruments.

Natasha, after the manner of young children, was inclined afterwards to be impatient with her mother, and her own memory being clear could not understand that her mother's loss of memory was something she could not control. "Don't you remember, Mummy?" she would say. "You were there. You were sitting in a chair, and Tanya was on your lap, and I was sitting next to you on another chair. Don't you remember, Mummy?"

It seems that at this stage, within the first week following the incident, only the two children had a clear recollection of what really happened. Later we shall understand why this is so. The younger child Tanya, who was only three years old at the time, was too young to be questioned independently; but I was satisfied from her reactions to the talk of her elders that she also had a clear recollection. Later, she began to volunteer remarks which made this quite clear.

The day following the incident, Natasha went to school, and mentioned to her teacher that her mummy and daddy had seen a flying saucer. The teacher did what any teacher would do in the circumstances, and told her to draw a picture, in coloured crayons, of the flying saucer. Later I saw this picture and photographed it.

The interesting thing about this picture is that it shows a feature – a broad double yellow beam diverging downwards towards the ground – which was not recalled by any of the adults until John saw it in his 'dream' a week later. Natasha's parents did not know she had spoken to the teacher, and did not know that she had made the picture, until the end of term much later, when Natasha brought her books home from school. So that John, when he dreamed about the yellow beam (see Chapter 2) did not know that Natasha had drawn a yellow beam coming down from the flying saucer; and Natasha, when she drew the picture, did not know about her father's 'dream' experience of a yellow beam because it had not yet happened. It is also interesting that this theme of the downward-diverging beam turns up again much later in the story, where it features as part of the design of the badge or insignia worn on their uniforms by the crew of the spaceship; clearly it has a symbolic significance for them.

It is also surely significant that Natasha, when they had arrived home and Frances was about to leave for her own home, warned Frances not to get 'sucked up' into a spaceship. Note that, in the 'cover story' which is all that the adults had conscious recollection of at that stage, nobody had knowingly been 'sucked up' into a spaceship. Her use of the term 'spaceship' is perhaps significant, because hitherto the grown-ups had used the term 'flying saucer'. As will appear presently, Natasha would have heard the craft they visited referred to by its own crew as a 'spaceship'; the Janos people, when speaking English, spoke of it as a spaceship, or more often, simply as a 'ship'.

It seems useful at this point to quote from the tape recording of a conversation which happened much later – on the first of October 1978 to be precise – when John, Frances and I had just been driving slowly along the section of road concerned with the story, checking locations etc. We had arrived at the entrance to Faringdon, at the place at which Frances and John felt they had 'switched back' into reality, and the 'cover story' and the 'real story' had abruptly reunited.

We had parked the car – almost exactly, as it later turned out, at the 'set-down point' at which the spaceship set them down back on the road – and we were talking over the incident. I quote from the tape:

JOHN: Well, we were discussing last night between ourselves, and we remembered that Natasha had made a remark, when Frances went to leave our house . . . to keep her windows shut and mind that she doesn't get sucked up inside a spaceship on the way home; and we were talking about this, and I realised and said to the others that it was something like that on our journey; that if they could imagine that after I got back in the car and we drove off, that we were as if sucked up inside something, and all sat in the car, imagining we were driving with like a picture screen at the cinema in front of the car, so that you could see hedges and bushes go by, and stayed in that position for the time that we lost, and were then dropped back down on the road without realising it, just as we came into Faringdon, and we saw the houses that we knew and were familiar with, at that point . . . because the road that we came along, as we said just now, certainly wasn't this piece of road here.

FRANK: [that is, myself, the author of this book] I don't think you were ever on that road at all.

JOHN: I know myself, though, I'm quite convinced that nowhere on that journey did we turn off or turn back on to a road. When we drove off, we stayed on that road. There was no possible turn. And we came back to reality, still on this piece of road, facing those houses.

What John said on that occasion is of interest in showing how, even at this early stage of the investigation, his mind was definitely turning towards the idea that they had actually been 'sucked up' in a spaceship, and away from the idea of the interminable closely-hedged road being a real experience. He did not get it quite right at that stage; but he was beginning to think of the experience of the 'cover story' being an artifact, a visual memory planted in their minds for a purpose. The 'real story' was beginning to take over from the 'cover story'.

It follows logically from this new pattern of thought, that the five people in our story had been for a time under the control of intelligent beings – the people of the spaceship. John at this time was leaning towards a realisation that the five of them had been actually in the spaceship. He speaks of being "dropped back down on the road" at the point where we were talking in the car, which is exactly what did actually happen.

CHAPTER TWO

John and Frances Dream

ABOUT A WEEK after the incident near Faringdon, John was taken ill with influenza, and was lying in bed in the daytime, feeling, as he says, "pretty rough"; and in this state, which was drowsy without actually being asleep, he began to recall experiences relating to an actual entry into the flying saucer by all the family. He thought of it at the time as a dream, and always referred, afterwards, to this part of his recall as a 'dream'; but as it were in quotation marks, as if he were not convinced of its being a dream, but felt that he had to call it a dream because, at that stage, he was still reluctant to accept the entry into the flying saucer as real experience; though later on he became convinced of it, and now is unshakable in his conviction. The same is true of Frances, who also had a 'dream' recall.

On several occasions, while he was describing this part of his recall to me, John laughed in a slightly embarrassed way, saying by way of qualification: "Well, of course, it was only a dream".

Now dreams can be very detailed; but there is always a recognisably dream-like quality about them, familiar to all of us. I will quote enough of the tape recording of John's account of his 'dream' recall to make it quite clear, I think, to the reader, that we are here dealing, not with a typical dream, but with impressions based on an actual memory. John, like all the adults, had a degree of amnesia about the entry into the flying saucer, which we now know was induced by the saucer people, to protect their visitors from the consequences of immediate publicity; they explained their purpose very clearly, as we shall see.

Amnesia can be relaxed by hypnosis, a technique which we

used extensively in the later stages of the investigation; my suggestion is that, in his abnormal state of mind as an influenza patient, John slipped accidentally into a self-induced hypnotic trance. The hypnotist who later helped us with the case advises me that this is perfectly possible.

I will now give a summary of John's 'dream' recall. It is necessary to be aware, at this stage, that while there is a large area of overlapping agreement with the later, much fuller recall under regressive hypnosis, there are some discrepancies between this 'dream' experience and the later hypnotic recall. As a matter of method, I always regard discrepancies as of particular importance; quite often they point the way to something highly significant: so I am never tempted to disregard them, and I shall bring them out clearly. The existence of a few discrepancies, on the other hand, does not in itself invalidate a statement; it indicates, quite often, that different parts of the experience are being described.

In his 'dream', John says that he and the others all got out of the car, and went up a sloping hazy beam of light into the flying saucer, entering through a doorway in the side of the ship. Through the doorway they found themselves in a corridor which extended to left and right, following the curve of the circular hull of the ship. The corridor was uniformly lit in a bright yellow colour. They turned to the left, where there were three doors to their right, leading into separate rooms. The two women, each carrying one of the children, were ahead of John in the corridor; John, obeying an inner compulsion in his mind, entered the first door, that nearest the entrance; and as he did so, he was aware that each of the women entered one of the remaining doors, each still carrying a child. (There is a discrepancy here, in that in the later, fuller recall under hypnotic regression, the two children both went with their mother Gloria, and Frances went in alone. Moreover, in the later account, the curving corridor was not entered directly from the outside, but was reached from the interior of the ship, which they had entered by another way.)

John says that he entered a darkened room, in which the only areas he could see clearly were a series of panels carrying instruments, extending round most of the walls. (This was not quite correct: the full recall describes instruments only on the wall opposite the chair, the side walls being blank.)

The instruments – meters with needles on a scale, switches, knobs, and small flashing coloured lights – were themselves illuminated, giving some general light over the floor; the panels had some thickness, standing a few inches out from the wall, presumably to allow room for circuitry, components etc. They extended from about table level up to an upper level about head-height from the floor, where the instrument panels terminated in a narrow horizontal shelf. The walls above this level, and the ceiling, were dark and obscure.

The room was roughly rectangular, its width being about ten to twelve feet, and its overall length about fifteen to eighteen feet. The right-hand far corner was dark and obscure from floor to ceiling; John had a feeling that someone was standing there, but could see no one.

In the middle of the floor, but nearer the door by which he entered, was a black-upholstered chair "like a dentist's chair", with arms and head-rest, the whole supported on a single stout metallic pedestal bolted to the floor. Something in his mind told him to sit on the chair; and he did so. John says that it was like a man's voice in his mind; he is clear that it was not a sound coming through his ears, but it had the quality of a sound, and he is sure that it was a man's voice and not a woman's voice.

(In the later, fuller recall under hypnosis, there were two people present who examined him medically; but they were both young women, clearly seen by John. A man had first ushered him into the room, who told him to wait for someone who would come to examine him; this man may well have told him to sit down, before the women came in.)

John, for some reason, associated the man's voice in his mind with the obscure corner on the right, and kept looking into this corner to try to identify the source of the voice; but he could see nothing.

In his 'dream' – but not in the later full recall – as soon as John had sat on the chair, it seemed to him that "something" came out from the wall on his right, and gripped him round the right leg, just below the knee. The grip was firm, but not uncomfortably tight. It was too dark to see much detail of what it was that gripped him, or what it was made of.

As soon as the thing had gripped his leg, a narrow beam of white light shot out from a source on the instrument panel directly in front of him, and shone on to his body. John says that where it struck his body, the circle of light was about an inch and a half across. The beam diverged only slightly as it came from its source, some ten to twelve feet away; it was almost a parallel beam, of a pure white colour.

The white beam scanned over him, starting on the left hand side, first slowly up the left arm and then down the left leg; then it moved across the chest and repeated the scanning movement, exactly the same, on the right hand side. John says the beam did not rise above shoulder level, and never shone directly into his eyes. He felt no warmth, tingling or other sensation in his arms and legs where the beam moved over him; it was just like a torch beam, nothing more that he could feel.

John said: "I got the feeling that the beam was sending out something – some form of power of some description – which was flowing through your body and then going back via the thing which was gripping you round the leg." If this subjective impression is well founded (and it may have been something that John was told), it would indicate that the thing which gripped his leg was probably some form of electrode contact, though one efficient enough to operate through clothing; John is quite clear, both in the 'dream' and in later recall, that he did not at any time undress: he was normally clothed throughout, and Frances agrees with this.

The business of the white scanning beam does not turn up again at any stage of the later investigation, either under hypnosis or otherwise. The clamp on John's leg does recur, though not exactly as he described it in the dream; and of course it may well be connected with the discoloured mark below the knee which all three adults developed after the incident.

Frances does not recall either a leg clamp or a white scanning beam; this does not, of course, prove that she did not experience them, because there are still parts of her recall which are hazy or patchy, as is also true of John.

In John's 'dream' sequence (but, again, not in any later recall), as soon as the white beam had completed its scanning movement over his limbs, it switched off; and at once the clamp around his right leg disengaged itself and retracted into the wall. He then got up from the chair and went out of the room by the same doorway he had entered by, coming into the corridor to find Gloria, Frances and the children emerging from the other doors. They all met together without speaking, and walked back along the corridor to the entrance door, returning to the car by way of the beam.

This return to the car direct from the medical rooms is completely at variance with the later detailed recall, whether hypnotic or normal. In the later recall, both John and Frances first visited other parts of the ship; and the five eventually all met together in a room on an upper deck.

It is clear, also, that the business of the scanning beam could only have occupied a minute or two at most; whereas there was a lost fifty-five minutes to account for. The 'dentist's chair' does reappear in all subsequent recall, and the instrument panels: the scanning beam and the clamp could well have been part of a real experience; but the way in which John's 'dream' has the five come out of the spaceship after only a brief examination, by a door directly to the outside, and without meeting any of the ship's company face to face, cannot be reconciled with the full and detailed story as we now know it to have happened, and as it will be described in later parts of this book.

(It is perhaps worth noting that an ordinary vertical door opening directly to the outside would be a physical impossibility, because of the shape of the hull; moreover it would provide no airlock.)

It is difficult to understand how this major discrepancy arose: possibly John's mind was rationalising from the need to explain how they got out of the spaceship again. This is perhaps the most unsatisfactory 'loose end' of the whole enquiry.

One detail from John's 'dream' does give an exact correspondence with what was normally recalled later on: the shape of the doorways, which John described as like an ordinary doorway, except that the upper corners were rounded, and the top of the doorway was slightly arched. On the other hand, there was a very odd feature about all the spaceship doorways, to which we shall refer later on; John's 'dream' does not mention this peculiarity.

Comparing John's 'dream' with his later, much fuller recall of the whole adventure, we have thus a mixture of things that fit and things that don't.

* * *

A few days later, Frances had a dream. This time, it behaved like an ordinary dream – it was at night, while she was sleeping. She was aware of John having had a 'dream', but only of the barest outline – he told her very little detail. Nevertheless, there are areas of agreement in detail between the two dreams. Frances explains that at that stage, the family did not think anything had really happened to them, beyond seeing a flying saucer with coloured lights, quite near to them; and they were inclined to joke about their dreams, rather than take them seriously. Frances, for instance, boasted jokingly to her brother John that her dreams were better than his: she saw men in her dream, whereas he had not seen anyone. Like most people, Frances and John were inclined to dismiss all dreams as being something unreal and fanciful, the work of an over-active imagination. They did not, at this time, believe that their dreams derived from memories of a real experience.

Frances was in normal health, and had been asleep, she says, for two or three hours. She half-woke, aware that she had been dreaming of going into the spaceship; and she told herself, as she turned over, that she must be sure to remember this dream, to tell John about it in the morning. Then, of course, she fell asleep again. Fortunately she did remember, on waking in the morning, quite a lot of detail; though she thinks there was a further part to her dream which she lost.

Frances told me about her dream much later; we had been discussing how they got into the spaceship, which at that stage was not clear:

FRANK: It seems most likely that you would go in together. Now, John's 'dream' version does see you all together; and he's got a most detailed description of exactly how you were all arranged, with John there, and you there on his right in the front, and Gloria in the middle behind you; you were carrying one of the children and Gloria was carrying another one, although he was not sure which was which. And that's how he sees you going up.

FRANCES: The dream I had was exactly the same in that aspect. It seemed to be like on an escalator; and I was standing slightly behind John, but more or less next to him, on his right hand side; and Gloria was just behind me, almost like a step lower if you were on an escalator; and she had Tanya, and I had Natasha.

FRANK: You were carrying Natasha?

FRANCES: Yes. And when I told John this, he seemed to think it was more than coincidence. Because although I knew about his dream, he hadn't – he just told me that we had all gone up together; he'd never told any of us in what order we were.

FRANK: The arrangement?

FRANCES: Yes. And he hadn't said who held Natasha and who held Tanya; but when I told him, he said that was exactly as he had dreamed it.

FRANK: Now, can you follow that further on, your dream sequence?

FRANCES: Well, as far as the dream went, it's just that we went up, and it was like going up an escalator, except that it was misty. We went through an opening in the bottom of the spaceship; and we seemed to be in a round room: we just seemed to come up through the floor. And there was a balcony going all the way round [this is not quite correct; the balcony was in two sections, each quite short] and I could see men on the balcony. The next thing I knew, we were going up another – well, it seemed to be like a slope that was moving, rather than an escalator.

FRANK: A ramp?

FRANCES: Yes; but it seemed to be moving.

FRANK: But did you – you didn't have to walk; it carried you up?

FRANCES: No; we just stood on it, and we went up. When we stepped on it, it moved.

FRANK: So, let's go back a bit. In your dream, you felt you came in –

FRANCES: – from underneath. If you had a bowl shape, and at the bottom of it a trapdoor opened, and we seemed to come through that.

FRANK: Right in the middle?

FRANCES: Right in the centre.

FRANK: What gave you the feeling of a bowl shape?

FRANCES: Well, it seemed to be flat on the bottom, and we came up through the flat bit, in the bottom. It seemed to be flat all around us; but it curved up to where the balcony was. [She is describing the 'engine room' of the spaceship, as we later came to know it; this is a big circular room with a floor which curves gently up towards its outer edge, doubtless following the curve of the saucer-shaped hull. As far as we know, the way up through the centre of the circular floor, through the airlock, is the only entrance to the spaceship.]

FRANK: Could you describe where you came in?

FRANCES: Well, it seemed to be a sort of opening here, which – I had a feeling in the dream – it was like two pieces of metal that opened up, and we came through here. And then I don't know where our feet were, but it seemed to close under us like that [a gesture of two things coming together rather quickly, in a horizontal plane].

FRANK: You had a feeling of sliding doors moving apart?

FRANCES: Sliding doors, two; yes.

FRANK: And you went up?

FRANCES: Yes.

FRANK: And then it closed. Was it a quick movement?

FRANCES: Yes; but not too quick to see. We were in the centre of a circle; the walls curved like this. And there was the ramp that we went up, to the balcony.

FRANK: Did the balcony have railings?

FRANCES: Well, it had a handrail on the top, like that; but below, I'm not sure if it was filled in completely, or barred. There were people here, two or three. And someone took us from where we came in, to the ramp – it was quite a long way – and pointed to it, for us to go up. We went up, and the two men were there, and some others on the other side. As we stepped on to the balcony, the men drew back to give us room. I put Natasha down, because she was getting

heavy to carry – and I don't remember any more. But I feel there was more, that it went on; but that's all I can remember.
FRANK: And the people you saw: can you say what they looked like?
FRANCES: No, I couldn't distinguish them at all, except that I knew they were in close-fitting suits; and they seemed to be silver in the dream: but that's all I could say.

* * *

Natasha also had dreams. From time to time, during the weeks following their adventure, Natasha would come into her parents' bedroom in the early morning, saying she dreamed she was in the spaceship; that she was in a room by herself, and many of the spaceship people came in to look at her. She said she didn't like them, because their eyes were "funny" – though she did not explain exactly what she meant by that.

(Later on in the investigation, when the grown-ups were beginning to remember with the aid of hypnotic regression, they also had some difficulty at first in seeing the eyes of people: at first they were seen as just a dark space, not seeing the eyes at all; later this became a dark ring round the eyes, "as if they were tired" said Frances; and later still, the eyes were seen quite normally and without any darkness, as pale blue eyes, but otherwise no different from ours.)

Although at that stage Natasha found these dreams worrying, she later became clearer and much happier about the whole experience; by the time I questioned her, she was entirely normal in her reactions to the spaceship people, and was not in the least 'put off' by her recollections of them. On one later occasion, when she happened to enter the bathroom where her mother was shampooing her hair, and (upon advice) had covered her hair with aluminium kitchen foil to prevent it drying too quickly, Natasha laughed and said: "Mummy, you look just like one of the spaceship people" (many of them wore a close-fitting silver helmet, continuous with the suit, which covered the head entirely, coming up under the chin, showing only the face – the kind known as a 'balaclava' helmet).

There was therefore enough to suggest that something more had happened to them, beyond the mere sighting of a flying saucer; though the family were reluctant to follow these leads to their logical conclusion. They had, nevertheless, a strong feeling that there was more to the affair than the 'cover story' had revealed; and they had a strong desire to find out more, if they could get someone to help them. They had mentioned their adventure, as far as they remembered it, only to family and a few close friends; but no one was able to advise them. Finally they went to the public library, which gave them the telephone number of the 'UFO hotline' for Southern England, as noted at the end of this book.

Ken Phillips, who has that number, passed the enquiry on to Jenny Randles, Secretary of UFOIN, who telephoned me, asking if I would like to take on the investigation, saying that it looked, from what she had been told, as if it had all the marks of a classic CE4 – a 'close encounter of the fourth kind', technical jargon for an encounter in which Earth people actually enter a flying saucer or spaceship, and have some personal contact with the crew. Such are still rare, though they are increasing in number.

I had a first general meeting with the three adults, who told me what I have related in chapter 1. Soon afterwards, John, Frances and I went by car to Faringdon, and drove twice slowly along the route which they should normally have taken, paying close attention to detail in the section with which the story is concerned.

The 'house that wasn't there' was still invisible and, indeed, non-existent; and we could not find any trace of anything which could account for the strange, interminable, narrow, closely-hedged lane. (Very much later in the investigation, we did find what we think is the physical basis of this part of the story; I will describe this when we come to it.)

CHAPTER THREE

Natasha's Testimony

THE ELDER CHILD, Natasha, was still only five years old at the time of the close encounter; but for a combination of reasons she is a particularly valuable witness. She is not old enough to dissemble convincingly, if one may be permitted a somewhat cynical observation about the human race; and though, like other people of any age, she is capable of romancing and embroidering the truth, it is immediately apparent to anyone who knows her, when this is happening. You can, so to speak, see the wheels going round in her mind. She can, of course, be mistaken; but so can any adult witness.

On the positive side, Natasha is an unusually bright child and very observant; and she has a good memory for detail. Her greatest advantage, in this investigation, is that she has had very little amnesia: she explains this by saying that, before they left the spaceship, the grown-ups were given a fizzy drink in a glass, "to help them to forget". This was later confirmed by Frances under hypnosis. Natasha refused the drink, and her memory is clear; it has, however, come back in stages, in a way characteristic of a person recovering from a slight amnesia.

I interviewed Natasha, by then aged six, on 21 February 1979, in the presence of her parents. Her younger sister Tanya, aged three at the time of the close encounter, sat beside her; and I noticed that, whenever Natasha hesitated over an answer to my questions, Tanya whispered in her ear and prompted her. It was quite clear to me, in fact, that Tanya, despite her extreme youth, understood all the questions, and knew the answers to them. (Since that time, Tanya has demonstrated that she had a very good recollection of the incident.)

I think we often under-estimate young children; many people dismiss children's evidence as unreliable, saying in effect: "Oh, well, it's only a child; you can't take any notice of that." In my own experience, young children are more reliable, as witnesses, than most adults; and I think most people with legal experience would say the same. Children are, of course, limited by what they know: an adult may be able to give much more information about a given incident than a child could, because the adult has a much wider background of knowledge to relate the experience to; by the same token, an adult is more prone to be misled by his own preconceptions.

In this instance, Natasha's evidence was clear and unequivocal, and was of great value to the investigation. In particular, she sorted out a confusion over the stopping of the car, by stating categorically that 'Daddy' (John) had stopped twice: that the first time, "he got out to look at the flying saucer; then, because Mummy was frightened, he got back in and he drived on"; but that a little further on, he stopped again, because the spaceship was right over the car, low down; and this time they all got out, and they 'floated' up into the spaceship, up a beam of light. At this stage of the investigation, none of the adults had realised that the car stopped twice; and it cleared away a difficulty.

Natasha said that a 'lady' called Akilias took Gloria and the children to a room, where many of the spaceship people came in to look at the children; then Akilias took her to another room, where she was shown pictures on a television screen of a flying saucer landing on a planet. She refers to the flying saucer as having a retractable undercarriage: "It had got legs; and they could pull them in when they didn't want them". (This tallies with some reports published elsewhere.)

On a later occasion, Natasha enlarged on and corrected her statement, saying that she now remembered that her mother (Gloria) and the two children were first taken by the lady to a room where there was a black couch (this could have been the usual examination chair laid flat, as we know from the other accounts).

She says that her mother was made to lie on the couch; and that clamps, with wires trailing from them, were attached to her mother's legs below the knee. Natasha says that she saw no more of her mother's examination; because Akilias said she would take Natasha to another room to show her some pictures. Tanya began to follow them out; but Akilias stopped her, saying: "*Not Tanya; she is too young: but you are just the right age to see them*". So Tanya remained with her mother. Natasha says that when she left the room, her mother seemed to be asleep on the couch.

I will return later to the matter of what pictures Natasha saw on the screen; they raise some problems of great difficulty, and may be very important.

Because it may interest the reader, I will now give a somewhat condensed transcription from the tape recording of the original interview of 21 February 1979. It should be stated here that, up to the date of this interview, the parents had purposely avoided saying anything about the flying saucer incident in the presence of either of the children, for fear of disturbing them – quite unnecessarily, as it turned out. It was the fact that Natasha, unprompted, began talking to her parents in a quite natural way about the spaceship, and seemed quite calm and unworried about it, that persuaded the parents that it might be a good thing to suggest that I should interview her; when the proposal was put to Natasha, that Frank should ask her some questions about the flying saucer, she seemed quite keen on the idea. She turned out to be an excellent witness, within the limits of her powers of expression; at first she was rather shy, but soon gained confidence:

FRANK: Now, Natasha, I want you to think about when you were in the car, and you first saw the light. Maybe you were asleep at first; but when you woke up, and you heard people talking about something in the sky – do you remember that, when you saw a light?

NATASHA: A light came down, and we were sitting in the car, and we got out of the car, and you were in the light, and you go up into the spaceship.

FRANK: Yes. Now, can we take it slowly; because I want to understand everything that happened. You say Daddy got out of the car first. [Natasha nods agreement.] And then what was the next thing that happened?

NATASHA: Mummy said get back in.
FRANK: Right; and what did he do?
NATASHA: He got back in.
FRANK: And what did he do when he got back in the car: did he drive along, or did he stop where he was?
NATASHA: He drived along; then he stopped.
FRANK: You are sure? Did he drive far before he stopped again?
NATASHA: No; not very far.
FRANK: And what happened: did you all get out?
NATASHA: Just Daddy got out with Frances.
FRANK: Can you remember: did Frances get out Daddy's side of the car; or did she get out her side and go round? Did you notice?
NATASHA: The same way as Daddy. [Later, Frances confirmed this.]
FRANK: So what about you and Tanya; what did you do?
NATASHA: We got out in a minute, we did.
FRANK: You got out after them? Now, you were in the back, with Mummy and Tanya; Mummy was in the middle – is that right?
NATASHA: Yes.
FRANK: And where were you sitting?
NATASHA: I was sitting the side where the flying saucer was.
FRANK: Could you see the flying saucer through the window? What did it look like?
NATASHA: It was shaped like a saucer shape.
FRANK: Like a real saucer, you mean?
NATASHA: Yes. The windows were straight like that [she held her hands apart to indicate a horizontal line] and got squares on, and people in them.
FRANK: I see. Now, this is how you remember it then at the beginning, is it, with windows? Did you say they were square windows?
NATASHA: Yes; cause there's big shape, with lines across, and down.
FRANK: And did you say you saw people in the windows?
NATASHA: Yes.
FRANK: And you saw them looking through the windows? Do you think they were looking at you?

NATASHA: Don't know.
FRANK: Were they looking down at you?
NATASHA: Yes.
FRANK: And then you said in a minute, you and Mummy and Tanya all got out?
NATASHA: Yes.
FRANK: Now when you got out, where was the flying saucer then?
NATASHA: It had moved over our car.
FRANK: Now just a minute: when you were still sitting in the car, looking at it through the window, which way were you looking?
NATASHA: That way. [She points to her right.]
FRANK: Out of the window to the right; was it behind the trees?
NATASHA: Yes; it started landing there.
FRANK: It started going down? Did it go up as well as down?
NATASHA: Up.
FRANK: As well as down? Did you see it go like that, up and down?
NATASHA: Yes.
FRANK: Did it have lights at that time? [Natasha nods.] What sort of lights?
NATASHA: All different coloured round; and when the next one goes, the other one stops. [?]
FRANK: You mean there was a row of lights, coloured lights, like the lights on a Christmas tree?
NATASHA: Yes.
FRANK: And you say it went round? Did it go round that way – to left – or that way – to right?
NATASHA: That way. [She moves her hand from left to right.]
FRANK: From left to right. So when you and Mummy and Tanya got out of the car, were you being carried, or were you standing by yourself?
NATASHA: Standing by myself.
FRANK: Could you still see the lights, the coloured lights?
NATASHA: Yes.
FRANK: You could? Now, what happened next?
NATASHA: Then the lights stopped, and . . .
FRANK: The little coloured lights stopped?

NATASHA: Yes. And a bright light came down, brighter than the other one, on us; and . . . we went in the spaceship.
FRANK: Now, this bright light: where did it come from? What part of the spaceship?
NATASHA: In the middle of a little hole.
FRANK: Where: at the side?
NATASHA: No, not at the side; in the middle of the spaceship.
FRANK: You mean underneath?
NATASHA: Yes.
FRANK: What coloured light was it?
NATASHA: It was a bright white one, too bright for my eyes.
FRANK: Did it make you blink, like that?
NATASHA: Yes.
FRANK: And did it shine right on you?
NATASHA: Yes.
FRANK: And it shone on all of you as you stood there on the road? [She nods.] Before you went up into the spaceship, did you see anyone else about, any other people? I mean, apart from your own family?
NATASHA: Only one. [John later, in regressive hypnosis, counted seven silver-suited spaceship people on the ground, around the car.]
FRANK: On the ground, while you were standing by the car? Somebody came down, you think? But only one? Was it a man?
NATASHA: It was a lady.
FRANK: How do you know it was a lady?
NATASHA: I seed lady eyes.
FRANK: And how was this lady dressed?
NATASHA: She had a gold suit, with a gold belt, a gold helmet.
GLORIA: It might be silver when she says gold. [Natasha is not always clear about the difference between silver and gold, her parents told me afterwards; but one must remember that gold suits have been reported in other UFO incidents.]
FRANK: Was it shiny?
NATASHA: Yes.
FRANK: And this was on the ground, before you went in?
NATASHA: Yes.
FRANK: Did she speak to you?
NATASHA: No.

FRANK: How did you know that you had to go up into the spaceship? Did she show you?
NATASHA: Yes.
JOHN: What did she have in her hand?
NATASHA: She had one of those things that they open the doors and shut them.
JOHN: What did it look like?
NATASHA: It's black, with red buttons.
FRANK: How big was it?
NATASHA: That big. [Indicating about three by one and a half inches.]
FRANK: And did she press the buttons?
NATASHA: Yes.
FRANK: And what happened?
NATASHA: And then we went up in the spaceship.
FRANK: What did it feel like when you were going up?
NATASHA: It made me feel dizzy.
FRANK: But were you standing up as you went?
NATASHA: Yes.
FRANK: Like going up a lift in a shop: is that what it felt like?
NATASHA: Yes.
FRANK: Now, this light that came down: was that still shining on you all the time?
NATASHA: Yes.
FRANK: What colour did you say it was?
NATASHA: It was a white one.
JOHN: Didn't you say we went up, and then we stood on a little ledge, just before the doors opened?
NATASHA: Yes; and then up to the middle. We stepped on a stair in the middle.
FRANK: When you got part way up, you stepped on to something?
NATASHA: Yes.
FRANK: On to a ledge?
NATASHA: Yes.
FRANK: Was it a ledge big enough to hold all of you, all the people?
NATASHA: [very positively] Yes.
FRANK: So it was quite a big ledge. What did it feel like, to stand on? Did it feel steady; or was it wobbly?

NATASHA: Wobbly.
FRANK: Was there anything to get hold of, like railings or anything?
NATASHA: No. [We thought later that this may have been a platform within the airlock, where they waited for the inner hatch to open.]
FRANK: How did you get into the spaceship?
NATASHA: Well, the door opened, and we went through a passage way; cause the lady took us there, and we got in the ship – it's a funny ship, kind of funny.
FRANK: You say a door opened?
NATASHA: Yes.
FRANK: What kind of a door was it: was it just an ordinary door?
NATASHA: It was a square door; and it opened by the top.
JOHN: Did it slide, or did it open?
NATASHA: It slided.
FRANK: Which way did it slide?
NATASHA: It slided that way; and the other one slided that way. [She indicated left and right with her hands.]
FRANK: There were two doors?
NATASHA: Yes.
FRANK: So it was square when it was open? Right?
NATASHA: Yes.
FRANK: Was it a big door?
NATASHA: [very positively] Yes.
FRANK: I mean a very big door?
NATASHA: Yes. [The hatchway was about fifteen by twelve feet.]
FRANK: What did you see when you got through to the inside?
NATASHA: Well, I saw a great big screen, and some people; and I saw our car in it. And the road; and some more cars going by.
FRANK: Did you see these straight away, as soon as you got in; or did you see something else first?
NATASHA: I saw them straight away.
JOHN: Before Daddy went off, and Frances went off; we were still all there, were we?
NATASHA: Yes.
FRANK: What sort of a screen was it – like a television?

NATASHA: It's a big screen, bigger than our television; it didn't have wood round it, but it was a big screen.
FRANK: Was it in the wall?
NATASHA: Yes.
FRANK: Not in a box?
NATASHA: No.
FRANK: And what did you see in the picture?
NATASHA: The car; and some more cars going by.
FRANK: When you saw them, were you looking down at them?
NATASHA: Yes; looking down.
FRANK: From above?
NATASHA: Yes.
FRANK: So you saw your own car – Frances's car?
NATASHA: Yes.
FRANK: And you saw other cars going by?
NATASHA: Yes.
FRANK: Your car was standing by the road?
NATASHA: Yes.
FRANK: Were the lights on?
NATASHA: No. Cause the lady had turned them off, cause Daddy left them on when he got out; and both those cars went rushing cause they were scared of the flying saucer. They saw a light.
FRANK: So you saw all this from inside the flying saucer, on the screen?
NATASHA: Yes.
FRANK: What sort of a room was it, that you were in, when you first came into the flying saucer?
NATASHA: There was a lift where you walked on; and if you want to go up, we went up and we looked around there [actually a moving ramp leading to a balcony].
FRANK: You said there was a lift; where was it in the room – somewhere at the side, in a corner, or did you have to go through a door to it?
NATASHA: Through a door to it. [This was a real up-and-down lift or elevator; it comes later than the ramp.]
FRANK: Was it just like an ordinary lift, like you have in a shop?
NATASHA: It was different.
FRANK: What was different?

NATASHA: It didn't have carpet [in fact, it didn't have a floor either]. It didn't have presses; cause the lady had it: she pressed it.
FRANK: You mean it didn't have buttons to press?
NATASHA: Yes.
FRANK: So what did the lady do?
NATASHA: Well, she had that black thing; but when she pressed the button twice, we went down and where the room we was going in.
FRANK: Suppose that was the thing – this is a bit too big [a cassette case], it wasn't quite as big as that, you said. Now can you hold it in your hand as if you were the lady, and show me what she did?
NATASHA: Well, she pressed the button there, twice.
FRANK: The same button?
NATASHA: Yes; and the door opened; and we went down [more probably up] into the room what we had to sit in. They took Daddy into another room, and took Frances into another room, and we were together with our Mummy.
FRANK: Now before, you told us you were sitting in a chair, and Mummy was sitting in a chair beside you.
NATASHA: Yes.
FRANK: And Mummy was holding Tanya.
NATASHA: Yes.
FRANK: So what were the chairs like that you sat on?
NATASHA: A kind of black – black ones.
FRANK: Like this chair?
NATASHA: No; but they had long things and round on the bottom, like that.
FRANK: How many legs did it have?
NATASHA: It had none legs; it had a round metal, like a pole.
FRANK: And the pole was fixed to the floor, was it?
NATASHA: Yes.
FRANK: What was there to fix it to the floor?
NATASHA: Something round. There was two round things, and it could be tightened by something. [Presumably a central column, secured to the deck by a circular base-plate with bolt-heads tightened by a spanner: more probably there were three bolt-heads; this would then agree closely with Frances's description, which Natasha had not heard.]

Natasha's Testimony

FRANK: Now the part you sat on: what shape was it?
NATASHA: Well, it was shaped like that. [Hand gestures indicating a curved surface with raised sides; this also agrees.]
FRANK: Did you see other people sitting on chairs like that?
NATASHA: Yes.
FRANK: Many people?
NATASHA: Yes.
FRANK: Were they men or ladies?
NATASHA: Some of them were men, and some of them were ladies.
JOHN: Didn't you say some arms came out of the chair and went back?
NATASHA: Yes.
FRANK: Can you tell me about the arms coming out of the chair: what was it like?
NATASHA: Well, on the sides, where you're sitting, they came up slowly together, like that. [She indicates by hand gestures two things arching over to meet in the middle.]
FRANK: Did they come around you?
NATASHA: Yes.
FRANK: What were the arms like, black or shiny?
NATASHA: Shiny.
FRANK: And did they join in front?
NATASHA: Yes.
FRANK: So where did it hold you?
NATASHA: The tummy.
FRANK: So why do you think they did that?
NATASHA: I don't know; to stop you falling out, I think.
FRANK: Did you feel any movement?
NATASHA: Yes.
FRANK: What moved, exactly? [She hesitates.] Natasha, have you ever been in an aeroplane?
NATASHA: Yes.
FRANK: Well, when the aeroplane goes up into the sky, you feel it go up, don't you? Was it like that?
NATASHA: Yes.
FRANK: You felt the flying saucer go up in the air, did you? Was it the same sort of feeling?
NATASHA: Yes.

FRANK: When you are in the aeroplane, they tell you to fasten something so that you don't fall out of your chair – a belt.
NATASHA: Yes.
FRANK: But this wasn't a belt; it was a silver thing, was it?
NATASHA: Yes . . . There was a lady looking after us; and the lady had a thing and she pressed it.
FRANK: So you were in a room, and a lady was looking after you. What did she do?
NATASHA: She talked to us.
FRANK: What did she say to you?
NATASHA: She said she was going to fasten the metal things round us. She said we might be going up in a minute.
FRANK: Did she say why you were going to go up?
NATASHA: No.
FRANK: Now somewhere – was it in that room? – somebody showed you a television with pictures on it.
NATASHA: It was next door.
FRANK: Another room? So when did you go into that other room?
NATASHA: When it was flying – the flying saucer was flying. She took off the belts and took us to a screen. [Later she said only she herself was taken.]
FRANK: In the same room?
NATASHA: Different room.
FRANK: You went into a different room where there was a screen. Was it like this, in a box; or was it on the wall?
NATASHA: On the wall.
FRANK: A big one, bigger than this? [indicating the domestic television].
NATASHA: Yes.
FRANK: What pictures did you see?
NATASHA: I saw there was the same flying saucer; what a flying saucer looked like.
FRANK: You mean you saw a flying saucer in the picture, like the one you were in?
NATASHA: Yes.
JOHN: She said the other day, she saw a planet which got closer; and then she saw other flying saucers sat on the planet. [It emerged from further remarks of Natasha, on a

later occasion, that she saw a film of some length, showing several different planets, with flying saucers landed or flying over them. Those landed had their tripod undercarriages down; she saw one retract its undercarriage on lift-off.]

We may note here an interesting remark made by Natasha in mid-August 1979. She had been chattering quite a bit to her parents about the flying saucer; John asked her: "Would you mind meeting the flying saucer people again?"

Natasha was, by this time, quite unafraid of the spaceship people; so her parents were surprised when the question brought an unmistakable flicker of fear to her expression; she answered:

"Well, it would be all right if it was them; but it might be one of the others."

When they asked what she meant by this, Natasha answered: "Like some of them we saw on the planets – some of them are monsters."

At first she would not enlarge on what kind of monster, beyond saying that they were 'hairy'. Later, when she was more used to the idea, she lost her nervousness of the monsters, and told her parents that on one of the planets she had seen (of course it could have been a place on Janos), there had been a stream flowing between muddy banks; and beside the stream were several 'men' – like ordinary men but very big and strong, and naked.

Their bodies were entirely covered with long smooth shiny black hair, which covered the face also except around the eyes; the hands and feet, also, were bare, and seemed 'whitish' – of course this may have been dried mud, for the hairy folk were in and out of the water. She saw one of the men stoop down and scoop up water in his cupped palms, to drink.

Natasha said the people lived under the ground: the film went on to show her the interior of a cave or large burrow, where a family of them, men, women and children, were living. They had nothing in the cave, except that they slept on very rough straw mattresses.

There is a strong resemblance between Natasha's hairy men and the 'sasquatch' tradition among American Indians, which

2 Natasha's drawing of Phusantheas

THE ORIGINAL IS COLOURED GREEN

has recognised UFO associations. I shall discuss this more fully in Chapter 13.

I have hesitated to include this incident of the hairy folk in my published account, since it depends on the unsupported evidence of a child who, at the time of the encounter, was only five years old; however, I am encouraged to do so, particularly by Frances, whose considered opinion, knowing the child well, is that she did not, and indeed could not have made it up.

This is also Frances's opinion concerning a further account of the 'monster' pictures, which Natasha gave me herself on a later occasion: after seeing the film of the naked hairy people, Akilias showed her a series of still photographs, not moving pictures, on the same screen.

There were four that Natasha remembered: Akilias told her that they were inhabitants of other planets which her people had visited; and she taught Natasha to say the names of them: SAUNUS, VONASON, FAUN and PHUSANTHEAS. When her parents were writing the names down to tell me, they began to write the last one 'Fusantheas'; but Natasha corrected them, saying: no, it begins with a 'P'. None of the family were aware that this gives the word a distinctly Greek flavour.

The actual appearance of the four alien types, of which Natasha later made sketches, is rather puzzling; and I do not wish to attach too much weight to them; because, while being clearly not human, they do conform loosely to the general 'humanoid' pattern. I feel that Natasha, in common with most other people, was being influenced by the feeling that 'monsters' from outer space ought to conform to the humanoid arrangement of head, body, eyes, mouth, arms, legs and so on. I may be doing her an injustice; but it is for this reason that I refrain from publishing the sketches, other than that of Phusantheas.

Akilias told her that, of the four, only Phusantheas was friendly; the other three were unfriendly, and Faun is shown carrying a 'gun'. It is of interest that of the four, Phusantheas is the only one which is at all human-looking; it is represented as a small, goblin-like man with large ears, and green all over. A 'little green man'?

CHAPTER FOUR

Entry into the Spaceship

IT WAS APPARENT to me that a genuine CE4 had taken place; there were the dreams reported by John and Frances, and there was a loss of fifty-five minutes of time unaccounted for. There were Natasha's early dreams; her interview had of course not yet taken place – I have brought it in early in the book because it is non-hypnotic, and because it provides a good introduction – but there was already enough from Natasha to support the feeling we were all beginning to have, that the family had actually spent nearly an hour inside a flying saucer.

At this stage, however, John and Frances were still inclined not to attach much weight to the evidence from dreams; and Gloria was getting no recall at all of the spaceship interior. Throughout, Gloria's amnesia has been severe and persistent.

I therefore proposed to the adult witnesses that I should seek the advice and help of an experienced hypnotist. Fortunately, a very good one was available in Gloucester, conveniently near at hand – Geoffrey M'Cartney, who has a busy practice as hypnotherapist and consultant hypnotist. Geoff is a man of great experience and much professional skill; the investigation owes a great deal to his cooperation. He has worked extremely hard over the hypnotic sessions; people do not always realise that the strain on the hypnotist is much heavier than that on the patient; and some of our sessions were very long, sometimes as much as two and a quarter hours, needing stamina as well as unfailing patience.

At the same time, I must make it clear that we used regressive hypnosis, not as the sole channel of operation, but primarily as a means of breaking the amnesia, of starting the flow of memory, which, once begun in the hypnotist's consulting room, continued and developed in normal recall. It was our practice to follow up each hypnotic session, two or three evenings later, by a normal question-and-answer session without hypnosis; these 'recap' sessions, stimulated by the foregoing hypnotic session, were extremely fruitful; and in case any reader is impatient to tell me that it could all have come out of the mind of the hypnotist, I should point out that many of the most important pieces of information were obtained in normal recall, on occasions when the hypnotist was not present. Nevertheless, the use of regressive hypnosis was necessary to the investigation: perhaps, if we had had unlimited time and patience, we might possibly have managed without; but it was at the very least a great time-saver in turning on the taps of memory; and the use of regressive technique did a lot more than that: it enabled the witness to re-experience the adventure, taking each stage at leisure, so that, with skilled questioning, he or she was able to go back into time and have a second (or third, or fourth) look at each situation and describe it in detail, for the benefit of the investigator and his tape-recorder.

In giving the results of the witnesses' recall, whether under hypnosis or normally, I must necessarily condense a great deal, and present the reader with a consecutive narrative, as I have done, for example, in the prologue of this book; but I wish it to be understood by the reader that every word of this narrative is derived from statements made by witnesses, or is a clearly indicated necessary consequence of such statements. Whenever possible, I have quoted directly – through the full recall of John or Frances – from statements made by officers and technical personnel of the spaceship, speaking English. Verbatim quotations from spaceship people such as Anouxia, Uxiaulia and Serkilias are printed in *italics*; this device has been used only where we are sure that these were the actual words used, and not a paraphrase.

To give a full transcript throughout the investigation, covering more than seventy hours of recorded interview, would occupy several heavy volumes, and be very tedious for the reader. I will, nevertheless, continue to give enough selected and condensed transcript to let the witnesses speak for themselves, and allow the reader some insight into their personalities. The transcript material I use in this book has all been checked, and where necessary amended, by the witness concerned; Natasha's transcript material in the previous chapter was checked by her parents.

The distillation of readable narrative from a mass of recording tape, bringing together information on a given incident or topic from material often gathered from different sources and on different occasions, has presented me with a formidable task; in the later part of the book, I shall do my best to say what I think it all means, and how the story of the Janos people relates to and impinges upon our own terrestrial story. Nor must we shirk drawing moral and ethical conclusions from what we are able to learn about the Janos story: conclusions which will sometimes be uncomfortable.

Perhaps the most difficult aspect of the story, for the investigator-narrator, has been the entry into the spaceship of our family of five. No doubt because it was the first part of their adventure really to hit them as something totally outside their experience, all the witnesses show some confusion over the entry episode, and had difficulty in recalling it. Frances, in addition, has a clear but cryptic piece of recall which comes in sequence, in the right place to be an entry recall, but which stands on its own: we cannot see how it fits in, or relate it to any other incident.

She remembers that she was standing upright, her body being moved or carried by an external force. Her movement was complicated: she likened it to standing on an ascending escalator in a London tube station (subway to American readers) but with her body turned left so that she faced the left-hand wall, where in London a series of advertisements are generally displayed. She was aware, in a confused sort of way, of a series of visual patterns moving past her vision, down to the left; but too quickly to recall any detail. At the same time, the 'escalator' which carried her was not straight as in a subway station: she moved, right shoulder forward, along a spiral path – as if, she said, an escalator were also a spiral staircase. The 'picture' wall was the outer face of the spiral; her back was to the unseen centre. The whole movement was quick and confusing: she sensed that she made several turns of a fairly tight spiral; then the recall abruptly cut out, leaving her with nothing to relate it to. None of the other witnesses have any recollection which in the least resembles it. We have no reason to relate it to the entry into the spaceship, except that, in hypnotic regression, it comes in the right place.

Frances does have, now, a fairly circumstantial recall of the entry sequence, which is in broad agreement with John's and Natasha's. In relating it here, I am combining information from all three.

As soon as John stopped the car the second time, it was surrounded by a white fog or mist. It seems likely that this was a genuine mist of water droplets – it certainly looked and behaved like one, eddying about with the slight air movements – but created by the spaceship immediately overhead, no doubt to obscure the scene from chance observers.

The spaceship was now low down overhead; we do not know if it was actually standing on its tripod undercarriage, straddling the road, or whether it was hovering stationary; either way the centre of the bowl-shaped hull was perhaps fifty feet above the roadway.

Although the Janos people regard these 'flying saucers' (as we call them) as 'small' ships, they are of considerable size; we finally estimated this one to have an overall diameter of about three hundred and fifty feet, the more deeply curved centre portion of the underside being perhaps two hundred feet across. It carried a crew of fifty-odd.

If its tripod straddled the roadway, the very size of the spaceship would make it unrecognisable as such by a passing motorist; it would look like a motorway bridge in a patch of mist, to anyone who did not know the details of the road well. Natasha did actually see (in a screen fed by a downward-viewing camera) two cars go by, soon after they entered the spaceship; so far from their drivers being frightened by the UFO, as she supposed, I think it likely that they did not see it, or registered it merely as a bridge in a patch of mist.

John got out of the car, leaving the headlights on 'dip' and the engine running. As he did so, several shadowy human figures appeared out of the mist, surrounding the car. They wore silver suits which gleamed where they caught the light from the car. Only the upper half of their bodies could be seen;

no doubt the mist was settling more thickly on the roadway.

John stepped out into the road; he remembers some of the figures coming towards him and then passing him by, as they closed in on the car. (Later, he was able to remember the scene more clearly, and said there were seven of them altogether.)

Meanwhile, Frances had scrambled out across the driving seat, and joined John in the roadway. Seeing one of the silver figures approaching the rear right-hand passenger door she went to intervene, thinking the children might be frightened; but the silver-clad person reached in through the open driver's door, released the safety catch on the rear door, and opened it; the person, with Frances assisting, then helped Gloria to bring the children out. Neither John nor Frances experienced any apprehension; and the children were too sleepy to be worried. Gloria cannot remember this incident.

The five now stood in the roadway, beside the car, which was parked close in by a grassy bank. They could see the dull black bowl-shaped underside of the ship overhead, quite clearly; but only the centre part: the outer edges were lost in the mist. They did not see a tripod undercarriage; but if it was down, the mist would probably have obscured it, and the whole area was darkened by the ship's shadow. From where they stood, no lights could be seen on the hull – neither the original white and red navigation lights, nor the ring of coloured 'Christmas tree' lights.

Suddenly a bright, but not dazzling, white beam shone down from a point in the middle of the hull overhead. It made a bright circular patch on the roadway in front of them, about five feet in diameter. Soon the beam was adjusted, moving back towards them until they all stood within the beam. As this happened, Frances felt a lifting sensation, as if her body were trying to float up.

A few seconds later, they all started to float up off the ground, quite slowly, standing upright in a normal position, up towards the ship's hull overhead. It seemed quite a long way up. There was nothing under their feet; they are sure of this. (The details of this ascent do not quite tally with John's 'dream' account, in which he thought more of a sloping way, like an escalator, up a yellow beam. It may be that John, in his 'dream', was confusing separate parts of his actual experience; I have no doubt that the version of the ascent given above is the correct one.)

Just under the hull, they came to rest; Natasha says that they stood on a metal ledge, which felt 'wobbly'. The others remember waiting, but are not clear about the ledge or platform. I think it likely that this platform, on which they stood waiting for a door to open – this thought was clear in all their minds – was actually within the airlock, which they had already entered through the open outer hatchway; looking down, they would still see the roadway and the car, quite clearly. Natasha reports (though the others did not notice this) that she saw one of the silver-clad figures reach in through the open driver's door, and switch off the ignition and the lights. (Note the evidences of familiarity with normal car design; even the safety catch on the rear door was familiar.)

Presently, they saw the inner hatch doors sliding apart, over their heads; these are a pair of heavy metal rectangular shutters, placed horizontally, which slide apart leaving a big rectangular opening, about fifteen by twelve feet. The opening movement was smooth and fairly rapid.

Soon they found themselves floating up once more, through the inner hatchway; somehow they were moved forward, so that their feet rested on the solid deck. Frances noticed, out of the corner of her eye, the hatch doors closing quickly behind them.

They found themselves in the centre of a vast circular room; John, who is experienced with interior measurements, later estimated its overall diameter at 150 feet; from various other considerations, taken together, this cannot be far out. John at first thought of it as like an aircraft hangar, or a large indoor multi-storey car park – the latter analogy was suggested by the many cylindrical columns or pillars which rose from floor to ceiling.

The floor was mainly horizontal; but towards the outer perimeter, it curved gently up into a bowl-form, evidently to follow the curve of the outer hull. Indeed, this big circular room, which we came to call the 'engine room', must have occupied nearly all of the lower part of the spaceship hull, that part which is externally expressed as a bowl shape. There must, of course, have been a space between the flat deck and the outer hull, except peripherally; and later, John was taken down into this lower deck, the central part of which is occupied by the airlock. John's tour of the engine-room complex was so detailed that I will give it a chapter to itself.

In front of them, some distance away, was a balcony with a handrail, not right back against the outer wall, but corresponding to the circle where the floor began to curve up – the relations will be better understood by a study of the various diagrams of the engine room.

A sloping ramp, also with a handrail, led up to the balcony; the ramp and balcony formed an arc of a circle concentric with the outer wall, but smaller. There was a similar balcony and ramp diametrically opposite. Behind the balcony, there was a wall from floor to ceiling; there were doorways in it, both on the balcony level and at the main deck level. Thus at two sections of its circumference, they were unable to see all the way through to the curving floor and the outer wall. They did not, of course, take in all these details at once; but John's later conducted tour was most thorough, and they all returned to the engine room at the end of the visit.

A silver-clad figure met them at the entrance hatchway, and accompanied them to the foot of the ramp; it seemed "quite a distance". Arrived at the ramp, the figure indicated that they should step on to it; as they did so, the ramp surface began to move forward, so that they did not need to walk (we have such moving ramps, for example in airports; the only refinement here was the automatic start when a person stepped on).

On the balcony, waiting to receive them, were three or four other silver-suited men. Frances says their manner was exactly that of a host welcoming invited guests to his home. One of them made a short speech of welcome, in good clear English without any trace of foreign accent, though the phrasing is not quite idiomatic:

"Welcome to our ship. Please, you must not be afraid at all. We mean you no harm whatsoever. We are going to examine you first of all, to see if you are the same as us. Then we will answer any questions that you want to ask us; and we will show you over our ship; and when that is finished, we will replace you back in your car, exactly as if you had never stopped."

CHAPTER FIVE

Frances Examined

AFTER THIS, THE adults were taken to separate rooms, the children going with their mother.

Frances found herself in a circular room; later she corrected this, saying it was more egg-shaped, with the narrow end of the egg in front of her. As in her dream, there was the black-upholstered 'dentist's chair' in the middle; banks of instruments lined the walls. Someone told her to sit down; so she sat on the 'dentist's chair'. She was not afraid; a voice in her mind told her that they would not hurt her: but she was worried because she did not know where John was.

Frances felt herself pressed down into the chair; she says that she was not held by anything, but it felt as if a heavy man were sitting on top of her; or, as she said later, more exactly as if her own weight had doubled. In the same way, her head was pressed back into the upholstered headrest. (In other contexts, we have learnt that the flying saucer people have complete control over gravitation; normally they use this control to obtain a lifting or hovering action, or a controlled ascent or descent: but presumably it could also be used to increase weight, doubtless in this case with the object of holding the 'patient' firmly in the examination chair, so that the body did not move about while readings were being taken; Frances felt at one point that she was told to sit still.)

At first, she was dazzled by an intense white light which filled the room, so that she could not see properly. It was, she said, like a car coming straight towards you at night when the headlamps are on high beam. It made her eyes water; she actually wept tears because of the brightness.

Then, abruptly, the light went out, leaving the room in semi-darkness; now, as soon as her vision adjusted to the gloom, she could see much better. The instrument panels carried many little domes which kept flashing with light – red, green and white, in an irregular way; there seemed to be no pattern to it. It was these flashing lights which provided all the illumination there was.

Frances began to panic; not on her own account, but because she did not know where John was. The first time we took her through this sequence under hypnotic regression, she was clearly distressed: "I can't find John" she kept saying, over and over again. At this stage her amnesia was only just beginning to break down, and her memory was patchy and confused; later it became very clear. I think, on that first occasion, she confused the momentary panic in the examination room with an earlier moment of anxiety, while she was still in the car, and John got out and she lost sight of him briefly in the mist, while shadowy figures came around her.

As soon as she began to panic in the 'dentist's chair', a voice said in her mind: *"You must be calm"*; and a white glowing disc, with two blue lights in it, in the place of eyes if it had been a face, appeared right in front of her; it may have been a hologram, but from what happened to John it is more likely to have been a picture on a small screen which was swung down in front of her eyes.

Curiously, this apparition had a calming and reassuring effect on her mind, as no doubt was the intention. Under hypnosis, she first said it was a face; then, seeing it more clearly, she described it as above. No doubt it was a psychological symbol; possibly it suggested a mother's face as seen by a baby. Anyway, it worked: she calmed down, and the symbol vanished. A voice in her mind told her that John was in the next room, and that they would all be together again soon. The feeling of relief was very strong in her mind.

She now realised that there were two men in the room, fairly tall and of slim build, in the usual silver suits. At first they appeared not to take any notice of Frances, and kept their backs to her; in fact they were attending to the multiple banks of instruments, reading data and recording it on small hand-held tablets with buttons – very like a pocket calculator but slightly larger; the finger action was essentially that of a person pressing buttons on a calculator.

Frances Examined 55

Frances had a very definite feeling (perhaps someone said it to her) that it was the chair itself that was examining her; and that the technicians were merely recording "what the chair told them" through the instruments. She was also told, in some way that reached her mind as an idea, that the data were concerned with facts about her body – pulse, breathing and so on.

The Janos people seem greatly concerned about the physiological similarity between themselves and the terrestrial human race. The way I read it, in the light of what we now know about their Earth origin and their intention to return to Earth, they are anxious to know whether, in the many thousands of years since they left the Earth, their own physiology and ours have remained sufficiently alike for them to be able to live with us, on this planet. John was told: "*We wish to examine you, to see if we can adapt*".

Later, Frances was told that medically, they could find very little difference between them and us: only the pulse rate was a little higher in the Janos people in the ships; but they expected that, if they came to live on the Earth, this would adjust itself automatically, without their having to take any special measures to correct it. (This is reported speech; but like other reported quotations – as distinct from the italicised direct quotations – elsewhere in the book, it is close to the sense of the original.)

It seemed rather extraordinary that the spaceships should be fitted out with several medical rooms, each with its own technical staff, for the sole purpose – as we originally supposed – of conducting examinations of sample Earth people to check up on physiological similarity; it has since seemed to me more likely that these are the regular medical rooms provided for the welfare of the crew; and that the 'dentist's chair' and its associated instrumentation are able to carry out not only diagnosis but treatment as well. It is clear that the medical knowledge of the Janos people is extremely advanced, especially in the field of mental science; and this consideration alone should make us anxious to secure their help and teaching, in return for a place to live.

I asked Frances to tell me more about the coloured lights. She said they were little domes; the red ones were larger than the green and the white ones, about three inches in diameter and hemispherical in shape. The red colour was like that of the rear lights of a car; the green was a really bright lime green. The white and green domes were about one and three quarter inches in diameter. The lights would go on for a couple of seconds or so, then go out; a number of them would be on at the same time – perhaps two or three red ones at the same time, and a number of green and white ones. The lights were arranged in vertical rows; and there were no lights near the floor or near the ceiling; the instrument panels – there seemed to be a series of upright panels set close together, all around the walls – extended from about two feet to about six feet above the floor.

Frances says that at one point, one of the men pulled a long lever which was fixed close to the wall, pivoting from a point only a little above the floor. The lever was about three feet in length. The man gripped the lever at the top and pulled it forward until it made an angle of about 45 degrees from its original vertical position. At the top of the lever, there was a sprung hand-grip device, like the brake handle in a vintage car. Frances herself compared it to the long levers in an old-fashioned railway signal box, which she had once seen in a film.

She distinctly saw the man's hand grip and squeeze the handle at the top of the lever, before he was able to move it; she demonstrated the action to me. From the way Frances described the man bending his back and flexing his knees to move the lever (again she demonstrated the action), it seems that pulling the lever down called for considerable muscular effort. They operated the lever only once while Frances was there. It seems surprising that an electronic technology such as that of the Janos people should find it necessary to make use of a heavy mechanical lever of this size; with few exceptions, Janos seems to be characteristically a push-button technology. No doubt there was a reason for this exception.

At one stage, while Frances was on it, the 'dentist's chair' changed its form into a horizontal couch, so that Frances found herself lying on her back looking up at the dark ceiling. At once she was almost blinded by a powerful white light beam which shone down, straight into her eyes; instinctively she shut her eyes tight, but she could still see the light through her closed eyelids. Soon it was switched off, and the room went dark again, to her great relief; and presently the chair went back into the sitting position.

I asked her: "Is it possible to say how long this went on – how long were you sitting in the chair?"

FRANCES: It definitely wasn't just a few minutes; it seemed more like twenty minutes.

FRANK: And all this time, the two men were working on the instruments?

FRANCES: Yes.

FRANK: Did you get impatient?

FRANCES: I didn't get impatient; but I began to feel a bit – you know, when you sit in a chair in one position, for any length of time, it begins to irritate you and – I felt I wanted to get up and walk about, if only to have walked over to see what they were doing. I'm quite sure I would have made the effort to get up, if I hadn't been told to sit there.

FRANK: Had you any bodily sensations?

FRANCES: No; except I could feel my heart beating: I could feel a sort of 'thump, thump, thump'.

FRANK: Was your heart beating normally, or faster?

FRANCES: No, not fast; just normal. Because I thought to myself, well, there's nothing to worry about, because it's quite regular, you know; I remember thinking that at the time.

I asked her to describe the two men and their clothing. She said that they were both very fair with blue eyes; the yellow-blond hair was cut very short, "like an American crew-cut". They were clean-shaven, without any facial hair other than the eyelashes and eyebrows, which were thin and not easily seen. The hair at the side of the temples had short 'sideboards', trimmed straight across. The ears were normal, perhaps a little on the large side. The men, she said, were about John's height (six feet) and of slim build, as John himself is.

They were clothed in a single all-over garment, shining like silver, with no division at the waist. The garment fitted the body closely, but was not tight. There was no helmet. Their shoes were black, with very thick white spongy soles, which compressed as they walked. At least one of them, that she saw closer, had a white disc on his chest, with some kind of device on it which at the time she could not remember clearly. Later, we had the details of the badge. Although many of the crew were later described as wearing belts, with the badge fixed to the belt, these two had no belt, and the badge was on the chest.

FRANK: The one that came closer to you: was there anything about his face that was in the least abnormal or unusual, apart from the ears?
FRANCES: No, I didn't feel so; he didn't look at all like we think of as a foreign-looking face; it was more like an English or American type of face. He didn't have very high cheekbones or anything like that.
FRANK: He was just like anybody you might meet in the street?
FRANCES: Yes; but sort of English or American, rather than Italian or Eskimo, you know.
FRANK: The hair again: it was really a close trim? It was cut short?
FRANCES: Like a 'short back and sides' that men used to have. He had a crew-cut, you know, straight back from his forehead.
FRANK: In other words, the hair stood straight up?
FRANCES: Yes.
FRANK: Would you say the hands were gloved or bare?
FRANCES: Bare. [One of the officers was later described as wearing thin silver gloves.]
FRANK: And how did the uniform end at the neck?
FRANCES: It was a round neck; but it had another piece on it, like a –
FRANK: Like a collar?
FRANCES: Yes, but not rolled over. It was quite low down on his neck, and it had another piece, a thicker piece, on the edge of it. You know, when you have a round-necked jumper with a thicker piece at the neck opening; it was like that; only I've got an idea it was a different colour, that band.
FRANK: Thicker, was it? Did it stand out?

FRANCES: It seemed to be rolled out a little bit, yes; as if it was a bit thicker. And it seemed to be more white, the same as the background of the circle on his chest.
FRANK: But was there nothing corresponding to that at the wrists?
FRANCES: No; it just ended plain.

Eventually, one of the men came over to Frances and told her that her examination was complete, and that she could get up; they walked together to the door. It was then that she noticed how the thick spongy soles of his shoes compressed as he walked. At the door, he smiled at her as if saying goodbye; and she understood that she was to go by herself through the doorway. She was told that someone would meet her and take her along.

FRANK: When you say the man smiled, it was an absolutely normal smile? The way anybody might smile?
FRANCES: A very friendly smile, yes, as if we were friends; a friend's rather than a stranger's smile, you know. Really a smile of pleasure, to say thank you very much, goodbye.

The man who met her in the corridor was described by Frances as "a very big-built man", about six foot two or three inches tall. As far as she could tell, he appeared to be completely bald. His uniform had special markings: on each side of the chest, a white band started just below the shoulder line, and went straight down, tapering gradually to a point about elbow-height. She also noticed along each shoulder a line in the silver fabric which, to her, suggested a concealed zip fastener. He had a belt with a circular badge in the front, attached to the belt by an elaborate silver clasp; the belt itself was transversely ribbed, suggesting an elastic material, but it was not unduly tight.

The big man took her along a corridor which curved slowly to the right, following the circular shape of the ship's hull. That they were, in fact, just inside the hull is made clear by the fact that Frances saw, on her left, a row of circular portholes of thick glass; the corridor was lighted where they walked, and she could see only darkness through the portholes.

The corridor was of considerable length; and as they walked along without speech, Frances noticed that the lights in the ceiling, which came from recessed circular fittings about four inches in diameter, switched on automatically as they approached them; and looking back, she saw that, at a corresponding distance behind them, the lights were extinguished one by one. Thus the corridor was automatically illuminated whenever anyone was walking along it; but electricity was not needlessly consumed when the corridor was empty. Evidently there is a need to conserve energy in the spaceships.

They passed several doors on the right, then came to a door at which Frances was received by another man with different insignia, whom she recognised as one of those who welcomed them on the balcony – but not the one who made the speech. He was about John's height, and he had no belt; instead there was a plain white disc on his chest, about five and a half inches in diameter, with no design on it.

The big man left them; and Frances was taken by her new acquaintance into the room they had come to. It was quite a large room, of odd shape, basically four-sided: the door they entered by was in the middle of one longer wall which followed the slow curve of the corridor; the side walls were straight, perpendicular to the corridor wall and thus very slightly angled; and the opposite wall was curved much more sharply than the wall it faced.

Evidently this was a relaxation room used by the crew in their off-duty times: the lighting was subdued and restful. It reminded her of a café; there were a dozen circular tables, each about four feet across, arranged informally as a grouping, not in rows. Round each table were four chairs. Both tables and chairs were supported by a single central column of bright metal, attached to the deck by a thick circular plate secured by octagonal bolt-heads. The seats of the chairs were black, a curving saddle with raised sides, to follow the shape of the body; some of the chairs, those backing on to the corridor wall, had a backrest in addition, also black and curved to fit the body.

There were about twenty people in the room, sitting round the tables, mostly just talking; but some had drinks in clear

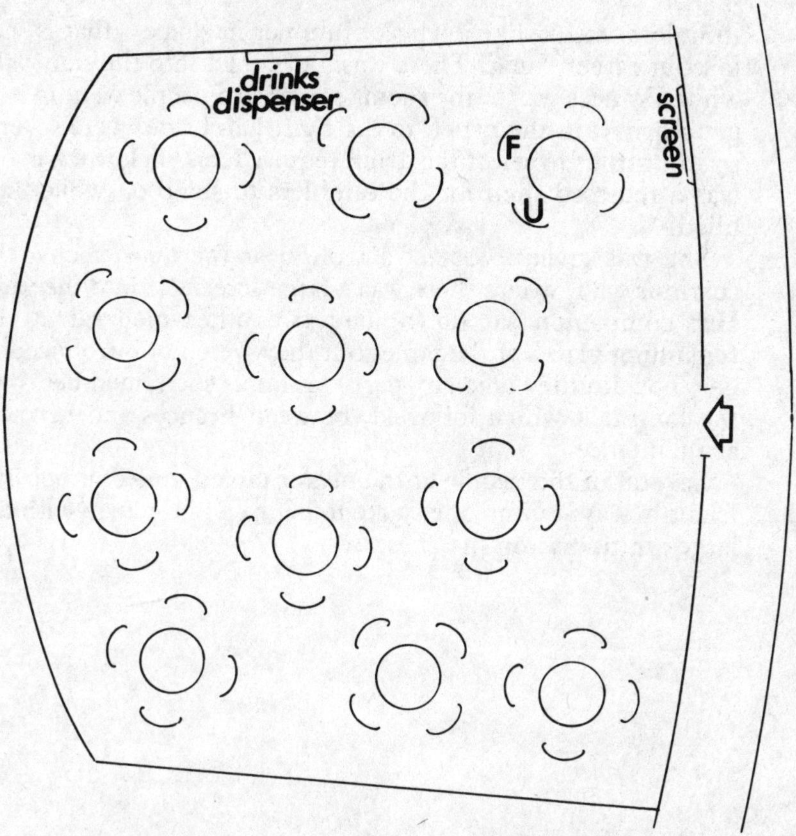

3 Layout of the room in which Frances and Uxiaulia talked

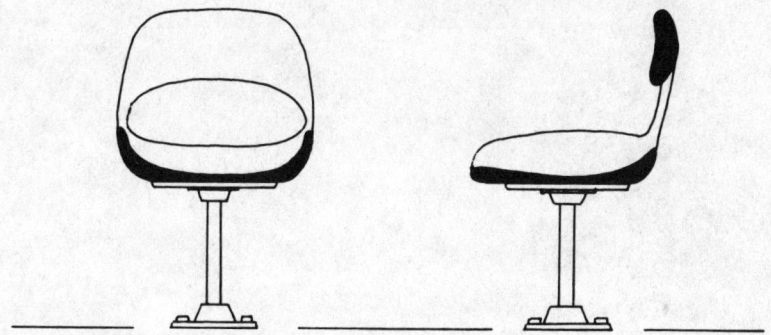

Detail of chair with back

tumblers, rather like a whisky tumbler in shape – that is, not tall but rather squat. There was a panel let into the side wall; while Frances was in the room, one or two people went to it to get a drink; in the panel, to the right-hand side, were several push-buttons to select the drink required, and below the panel was a recessed shelf for the tumblers to stand on while they filled.

She was given a seat at a table near the door, facing the corridor wall, where there was a large screen let into the wall. Her companion sat down; and two other men sat at the remaining places at the table, but they were not introduced to her, nor did they take any part – as far as she remembers – in the long talk which followed, between Frances and her new acquaintance.

Several of the people at the nearer tables smiled at her in a friendly way; but no one spoke to her except the man with the large white disc on his chest.

CHAPTER SIX

What Frances Was Told

THE MAN WITH the large white disc on his chest began by asking Frances her name. Then he said: *"My name is Uxiaulia"* – and he spelt it for her; a succession of letters appeared in her visual mind, one at a time, as dark capital letters on a light ground, with a timing of rather less than one second apart – UXIAULIA. He pronounced it carefully for her, making her repeat it until she got it right; it is pronounced as 'Yóuksia-óolia', with two equal stresses.

He said: *"I am a top explorer pilot; I am from Janos"* – and again he spelled it out for her, JANOS, and taught her to pronounce it, as 'Jáne-oss', the first syllable stressed. He encouraged Frances to ask questions.

Frances asked: "Where is your planet?" and he replied: *"Further away than you have ever dreamed of"*.

To the question, "How far away is it?" he answered: *"Several thousand light years"*.

Then Frances asked how long they had been travelling, and got the surprising answer: *"For two of your years"*.

When I first heard this answer, in a hypnotic session, I was puzzled and disappointed; for a travel time of two years would put Janos within the solar system, where we know very well there is no such planet.

Then I remembered that this would not necessarily be so if their velocity of travel had been so high as to approach at all closely to the speed of light. Relativity theory tells us that, under these conditions, there are two ways of measuring time, or rather two points of view from which to measure it: from the external point of view of an observer not moving at high speed (of course nothing in the universe is ever really at rest; but relative to a speeding starship, an observer sitting on a planet is not moving very quickly), or from the internal or subjective point of view of an observer riding in the starship.

At moderate speeds, time will flow much as it does for us; but if one is travelling at a speed close to the speed of light, one's own subjective time slows down quite a lot; if one could ever perform the impossible, and equal the speed of light, one's subjective time would come to a standstill. There is always some slowing of time for any body in motion; and all astronomical bodies are always in motion, so no one ever has quite the right time – it is always a bit slow. But it is only when velocities get near to the speed of light that the difference becomes noticeable. If a ship can travel really close to the speed of light, the difference in the rate of flow of time can become large; so that, as in the example given us by the Janos people, a journey which according to clocks and calendars on Janos or on Earth takes several thousand years could be travelled in only two Earth years, according to clocks and calendars inside the spaceship. Everything inside the ship would go by ship's time, including living processes; the people in the ship would end the journey only two years older than when they set out, thousands of years earlier by planet-time.

The Janos people are fond of a joke; so it may have been with a twinkle in his eye that Uxiaulia now put a question to Frances: *"How old do you think I am?"*

Frances, surprised, said: "Oh, about thirty, I suppose" – going by his looks and physical condition, though there is some indication that the Janos people do in any case keep their youth longer than we are yet able to, though we have made a lot of progress in that direction in the last few centuries.

Uxiaulia laughed at this, and said: *"I am really a very old man. We are all very old, because we have travelled so far"* – he emphasised 'very'. But he went on to explain that they had aged only two Earth years during the journey; so that, although the events which drove them from Janos happened thousands of years ago, it does not seem so long ago to them.

Uxiaulia said he would show Frances a film to explain what happened; why they had to leave Janos: he indicated the screen in front of her. The subdued lighting made it easy to see the pictures: all too easy, with the crystal-clear realism that is characteristic of Janos screen technology.

We have already heard of this film in the Prologue. As Frances re-experienced the film under hypnotic regression, her distress was sharply evident to myself and to the hypnotist. I think there is no doubt that, while Frances was herself greatly distressed by what she saw in the film – for she is a person endowed with much sympathy in the literal sense – she was also picking up the more personal distress of Uxiaulia by a kind of mental link-up.

So that the reader shall understand how this terrible story hit Frances, I will quote directly from a transcript of part of the hypnotic session of 5th March 1979. Frances is in deep trance: Geoff M'Cartney, the hypnotist, has taken her back in time to the exact point at which Uxiaulia switches on the film, as he and Frances sit at a table in the café-like room in the spaceship.

GEOFF: Where are you, Frances? What's happening?
FRANCES: [in distress] Looking at a screen . . . people . . .
GEOFF: Can you describe them to me? What's bothering you?
FRANCES: [in severe distress, her breathing laboured] They're sick.
GEOFF: What's wrong? Can you tell me?
FRANCES: They look like Oxfam people. [I think she is remembering a picture of lepers in an advanced stage of the disease. What she is seeing in the film is the part where people on Janos are slowly dying of radiation sickness.]
GEOFF: Where are they? What sort of background are they in?
FRANCES: [still in great distress] He said: '*My people . . . my people . . . my people . . . they are dying . . .*' Something crashed into them. From the sky. Their houses are on fire. People screaming. Rocks . . . meteorites. Dust. The dust made them sick: it's radiation. They had to leave them behind because they were too sick. I don't know . . . Ruins . . . [a long silence].
GEOFF: What's happening?
FRANCES: [easier] We were arguing. I said they should stay and help them. They said it was the only way they could survive. I said they should help. They said they must live

so that the same thing doesn't happen again. He said they're all dead now; they are not suffering any more. I said I'd rather have died with them. He said: *'Not everyone looks at it like that; we had to survive, to make sure it does not happen again'*.

In fact, Frances was under a misapprehension: she thought at first they had abandoned the dying people to their fate; whereas, as she learned in a later regression, going through the experience a second time, they did everything they could for the sick people, short of themselves becoming contaminated; and the fleet did not leave its orbit around Janos until the last of the radiation victims was dead.

The reference to 'meteorites' also calls for explanation. Uxiaulia, trying to explain the falling rocks to Frances, first called them '*boulders*'; then, sensing this to be inadequate, he seemed to be casting around in his mind for a stronger word that Frances would know: '*meteorites*': but Frances understood that they were not actually meteorites – this was merely a word he used by way of illustration. In fact, there are good reasons why they could not have been meteorites: because of its high velocity, a meteorite 'as big as a house' would explode on impact with enormous violence, blowing a crater miles across; whereas these rocks merely fell and bounced, leaving a land surface densely covered with rocks of all sizes, piled one on another, but with no suggestion of cratering. This was clearly shown in a film seen by John.

FRANCES: [continuing] They've got special dried food. They have to mix it with liquid. They've got a whole ship full of it. They've got a lot left. [Frances had the impression, as she explained later, that they were anxious that we should not think they were in urgent need of supplies; their need is for a planetary home. Sadly, it seems likely that the migration fleet was provisioned for a much larger number of people than eventually left Janos; so they have a surplus, and the big ships are not as full as they were planned to be.]

GEOFF: Why are they travelling? Why are you on board?

FRANCES: They wanted to see what sort of people we are. Want to see if they could live here. They can't go back . . .

What Frances Was Told 67

GEOFF: What else can you see on that screen?
FRANCES: We haven't been looking at the screen; it upset me too much, so he turned it off.
GEOFF: What is he telling you about?
FRANCES: He told me what happened: they left; they need somewhere to live; they can't float around in space for ever. They want to come here; but they don't want to cause trouble. They've been talking to different people to see what ordinary people are like. They wanted to see if we are medically the same as they are. There seems to be very little difference. They want to come and live here; but they are trying to make sure they can do it without causing any war. They said there can't be a war; they've had too much trouble. They would rather die than cause any war.
GEOFF: Why do they think it might cause a war?
FRANCES: Because people want to use their knowledge to control other people. [This seems a fair summary of the reported motivation of American and Russian military leaders, regarding information about UFOs.]
GEOFF: Have they said where they are going to stay, or what they are going to do?
FRANCES: They think they could come in very small groups now; but they want to stay together. There are enough of them to fill one of our large cities.
GEOFF: Is this an advance party?
FRANCES: They are people that have been chosen to make the first contacts.
GEOFF: Do you know where the other ships are?
FRANCES: In space, waiting.

Uxiaulia showed Frances other, more cheerful, films, of what life was like on Janos before the disaster. One of these sequences was of a lakeside barbecue party at dusk. He also showed her, on the same screen, a still photograph of his wife and two children, in front of their house; they were killed in the rockfall. One of the first rocks came straight down on the house, while he was away. Perhaps he had taken the photograph himself.

They sat for a long time, looking at pictures and talking about them. Much of the background story, of what happened on Janos, already told in the Prologue, comes out of this conversation, and we need not repeat it now. At one point, Uxiaulia, remembering his duties as host, asked Frances if she would like to be shown round the ship, and began to tell her how they stored energy in the bottom of the ship – in this context, Frances, under hypnosis, brought out with great emphasis the words STATIC ELECTRICITY. It is, of course, possible to store energy in this form; but terrestrial engineers would have difficulty in storing enough static to be of much practical use.

Frances declined the conducted tour, saying that she would really much rather just sit and talk. She herself was not very interested in machinery, except her own car – and she broke off to tell Uxiaulia that she was worried about her battery, because John had left the lights on. Uxiaulia told her at once that she had no need to worry, because they had turned them off. This agrees with Natasha's observation.

Frances went on to say that her brother John was very interested in machinery; at this, Uxiaulia smiled and said: *"That is being taken care of"*.

The lakeside barbecue scene – evidently a favourite recollection of their lost home – yielded much detail. We have noticed elsewhere the incredible realism and sharpness of definition in Janos screen pictures: presumably they are electronically generated; but they are very much clearer than our television pictures, and our witnesses have several times remarked that the pictures are so vivid and real that they feel that they themselves are actually present in the scene, and are experiencing reality. This is how they report events seen in a film. There is a full three-dimensional stereo effect, which of course helps realism. On the other hand, they do not always use sound, except where it is relevant.

The scene is at twilight in the evening, on the shores of a lake. Here and there are 'trees' bearing large yellow fruits, rather like a melon. They are not true trees, as we understand the word, since there is no trunk or bole; several main branches come directly from the ground. The foliage suggests a rubber tree; the leaves are the conventional leaf shape with a midrib and a slight point, but distinctly fleshy, and a deep bottle-green in colour. Coloured electric lights are fixed here and there to the higher branches, giving the scene a dim romantic illumination; though Frances says the pictures showed that some light – a very soft light like starlight, perhaps – was coming from the sky above.

The barbecue stove itself was a rectangular metal box, dark-coloured, about three feet high and eighteen inches square on plan. Towards the bottom of each face was a series of horizontal ventilating slots, to allow air intake. The heat was smokeless, and there was nothing to indicate how it was produced; there were no visible electrical leads. On the square open top, several long metal skewers lay across the box; each held a series of lumps of dark-coloured flesh cooking in the heat. Uxiaulia told her, about the *'meat'* as he called it: "*We get them from the rivers*".

A man, who wore only swim trunks, sensibly, because the job was a warm one, kept turning the skewers; as he did so, from time to time he poured an oily liquid over the food from a small frying pan, just an ordinary frying pan, which he balanced on one corner of the box, to keep it hot.

Several couples or small groups of people sat around on the ground, eating, talking and generally having fun. Frances says they ate the kebabs with their fingers; but she was not sure if they would do so at home: after all, this was only a picnic. Other people strolled about in company.

Some were eating the melon-like fruits: of course they were not really melons; that was what Frances compared them to. They were large egg-shaped fruits, with a skin of a dull mustard-yellow colour; there was a great pile of them on a big shallow bowl, about three feet across, of a dull metal that looked like pewter. These yellow fruits seemed to be popular.

Some of the men wore swim trunks; others had something like a track suit – one of these, a red suit, had a white stripe down each outer side of the sleeves, and also of the legs. It had a broad waist band joining the top to the trousers.

Most of the women wore full-length skirts, nearly to the ground, with a long-sleeved, round-necked bodice. The material was filmy, like a chiffon or a fine nylon: most had white as the background colour; variation was obtained by coloured patterns printed on the skirt material, the bodice being plain.

The skirt, in all examples seen, was draped in a series of overlapping folds, cascading down the left side, secured by a metal clasp on the right hip. (Clearly we are dealing with a fashion.)

The printed pattern was made by a loose, open repetition of a large abstract floral motif: Frances says it was not really a flower – just the minimum conception of a flower, expressed in a few curved lines. The metallic-looking clasp was shaped to echo the same abstract floral form.

All the women Frances saw on the lake shore, in the fairly restricted area covered by the camera, wore dresses which were variants on this theme; one woman, quite near the camera, wore a striking ensemble in black and white: the bodice was black; and the skirt, with the usual cascade of drapes down the left side, was white, with black floral motifs; the clasp also was black. Her head was covered with something, also black, which Frances could not quite make out the details of. One is tempted to wonder whether the black head-covering, with the use of black elsewhere in place of a colour, may have been a sign of mourning; but really we have no evidence for this.

On the lake itself, the background to the barbecue scene, Frances saw two boats go by. These were quite small, perhaps eighteen to twenty feet in length, holding a crew of two, both seated amidships, one behind the other. They were driven by engines powerful enough to raise a considerable bow wave, with the usual patterns of ripples fanning out towards the shore. The bows, instead of rising forward out of the water, as with nearly all Earth boats, sloped down forward into the water. The stern, on the other hand, showed its underside clear of the water; under the overhanging stern, the water was turbulent as if driven by a propeller or other device having a similar effect.

The midships section was open, with seats for the crew; but the bow and stern sections were closed in above: a low windscreen, curved and raked, rose up at the front of the open midships section. The effect was not unlike that of a sports car, translated into boat terms.

Towards the bows of each boat was fixed a short staff flying a pennant flag, a triangle about twice as long as its hoist. The ground colour of the flag was deep blue; upon this background was a white disc, almost touching the hoist and the other two sides of the triangle; upon the disc was a device of a narrow deep blue line looped over itself, with a circular spot in the same deep blue over the point at which the looped line crosses over itself.

Each of the two boats carried a similar pennant, with one small difference: that on the nearer boat ended normally in a single point; but the other pennant ended in a double-pointed fishtail. The two pennants fluttered briskly in the breeze created by the boats' speed.

The nearer boat's hull was painted a pale blue; the one further from the camera was bright red. The inside of the open midships section, where the people sat, was white.

The two people in the nearer boat were dressed alike, in a red track-suit, as far as the upper body could be seen. A woman was in front, apparently at the controls; a man was the passenger behind her. Women and men on Janos, even when dressed alike, were readily distinguishable; all the women wore their hair fairly long, in the same 'page-boy' style already seen in a few women in the spaceship – though the majority in ship's uniform wore helmets: whereas the men all had their hair very short and neatly trimmed.

All the people in the lakeside scene appeared to be enjoying themselves; there was an atmosphere of relaxation and happiness.

Other dress details are given by Frances from the still photograph, shown on the same screen as the films, of the young wife and children of Uxiaulia in front of their house.

The girl, about five years old, was dressed in workmanlike red dungarees over a long-sleeved white jumper with a high round neck; the straps of the dungarees were secured, in front of each shoulder, by a white circular button or buckle.

The little girl had curly yellowy-flaxen hair; most of the hair was brushed out free, but on each side of the head, a bunch of hair was secured by a circular red hair-slide or grip, so that the two bunches stood out to the sides.

The boy (about three years old by his appearance) was

4   Pleasure boats on a Janos lake

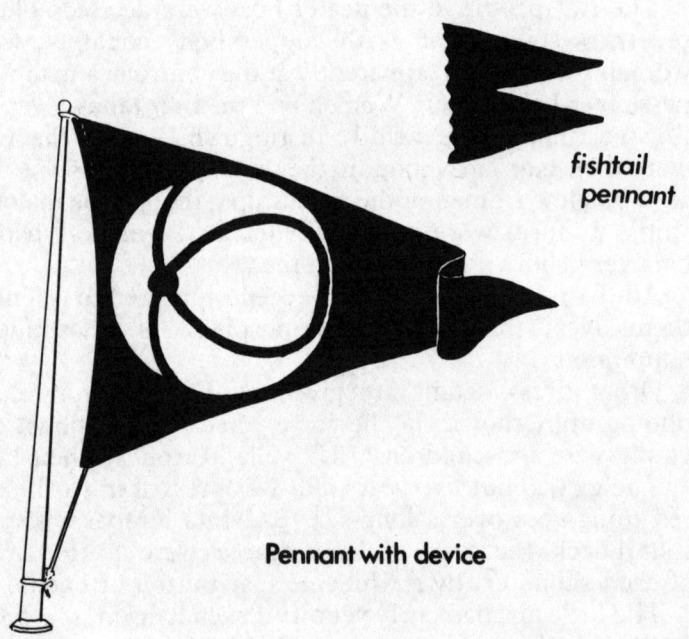

fishtail pennant

Pennant with device

dressed in the same way as his sister, except that his dungarees were pale blue. Both the children wore white shoes; the shoes were not open sandals, but covered the foot completely: Frances could not quite make out how they were fastened.

The mother (Uxiaulia gave her name as Vurna,* and her age as twenty-three – whether this means Earth years or Janos years, we do not know) was dressed in the same way as her children, in red dungarees over a white jumper; but her sleeves were shorter, ending above the elbow; and her shoes were red. Her fair hair was cut to about chin level, and was curled under at the ends.

The house and garden in the photograph are of interest. The house was seen cornerwise. It was of simple construction: a rectangular symmetrical plan with a simple pitched roof, the ridge parallel to the house frontage as usual. The front wall had a central doorway, with a window at each side. The doorway was an arched opening in the wall of the house; the impression was that it was just an opening, leading to an internal porch, with doubtless an inner door not seen.

The two windows were alike: each consisted of a large rectangular sheet of glass or other transparent material, curved into a bow which projected from the line of the wall. Frances could not see into the house interior.

The walls gave Frances the impression of being made of wood: certainly they were constructed of horizontal planks of some wood-like material, finished white, with the joints quite apparent.

The end wall was blind, also of white horizontal planks; attached to the roof gable was a bargeboard, decorated by a series of rectangular recesses carved into the timber. There was a carved boss or finial at the junction of ridge and gable.

The roof was covered with square grey tiles; but instead of overlapping as our tiles do on a roof, these squares all lay in the same plane, with no overlap. Evidently the joints, which again were quite apparent, were waterproof enough to keep out the rain. There was no chimney or ridge vent of any kind.

Surrounding the garden was a low white fence consisting of

---

* pronounced Voorna

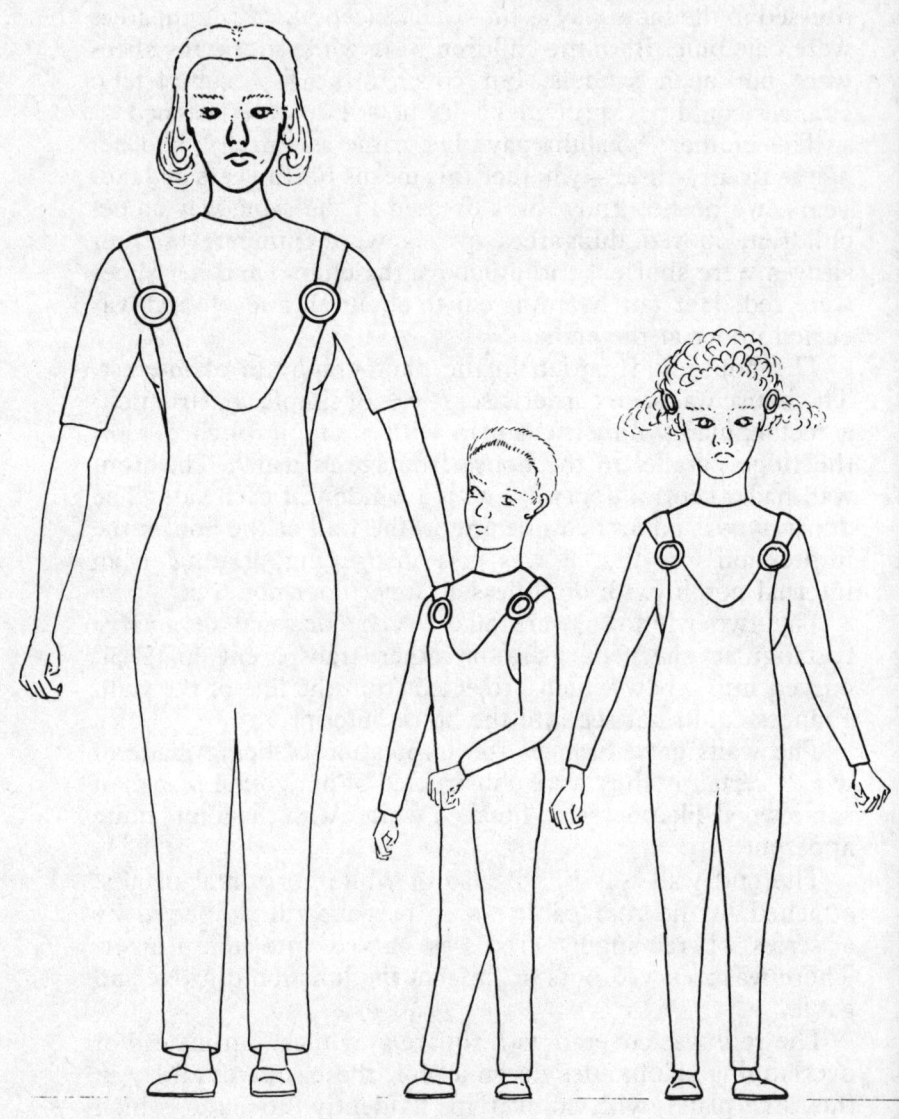

5 Domestic dress and hair styles for Janos young woman and children

short posts with a single horizontal rail half-way up them. The small 'garden' itself would not score many points in an English village. The surface of the ground was covered by a fine white sand, from which, here and there, bunches of spear-like grass blades emerged, of the usual deep bottle-green of the Janos foliage. There was just one flowering shrub seen in the picture; some of the flowers were red, and others pink. The shrub stood close to one corner of the house.

Frances remarked that in a similar photograph of an English house, you would expect to see in the background some other houses, perhaps a tree or so, and very probably a road with cars. In this picture, she was not aware of any background.

She formed the impression, in fact – perhaps from what she was told – that the house was close by the place where the film was made of the lakeside barbecue; the film may well have been an amateur one, shot by Uxiaulia himself with his own ciné-camera, or whatever is their technical equivalent. She was not aware of any sound with the barbecue film.

It occurred to her – and again, this could have been an idea given to her – that this house was no more than a beach-hut, for week-ends or holidays; however, we have no certainty about this: the house, though modest, would have been adequate for a couple with two young children, especially if the husband were away a good deal.

It was, in fact, the tufts of grass poking up through the sand that gave Frances the impression of a seaside or lakeside dwelling; she said it was like you see sometimes by the sea.

One thing struck me as odd, in relation to the Janos people's highly-developed technology: this was the extreme simplicity of their colour schemes; they seemed to use very few colours. Their clothing, on Janos, was simple and practical, except for the rather elaborate gowns worn by many of the women for an evening social occasion. We notice, however, that a woman in a boat wore a track-suit like a man's.

Despite their advanced technology and generous leisure time, the Janos people do not give us the impression of going in for sophisticated luxury. Frances had noticed that several of the women she saw in the café-like room wore make-up to darken their eyelashes; this was one of the few examples seen of what may be called vanity. Incidentally, eye cosmetics are one of the very oldest of all technological developments in Earth's history; some scientists think that the discovery of metals, copper especially, happened accidentally in the course of manufacturing eye pigments, back in the neolithic; and that this was the forerunner of the bronze age.

One isolated point which will interest many readers: Frances, thinking of the wolf-dog mentioned in the Prologue, asked Uxiaulia whether the Janos people kept pets.

"*No*", he replied; "*we do not keep animals as pets indoors, as you do*" – and something in his expression told Frances that he thought it a rather odd thing to do. "*We have animals for food*" he went on; "*we do eat some meat; but mostly we eat the things that grow*". Frances was not able to say, whether he meant by this that they grew crops, or that they depended on things which grew naturally, such as the melon-like fruit.

At one stage in the film – she is not sure exactly where it fits in – she was on a hillside, in daylight, looking across a small valley to a hillside opposite. Filling the valley, and spreading up the opposite hill, were many of the 'bungalows' – single storey houses – similar to the one in the photograph. From her elevated viewpoint, Frances was mainly conscious of the pattern made by many roofs; she was not aware of roads or traffic.

Indeed, it is quite possible that Janos did not need access roads, if their transport was off the ground, as we have seen in some examples. Uxiaulia remarked at one point: "*Our transport is different from yours; our cars float above the ground*".

Uxiaulia and Frances talked for a long time, probably getting on for half an hour. Their talk might have gone on longer; but a woman with long hair came in and spoke to Uxiaulia in their own language. She said something also to Frances in English, which unfortunately she cannot remember. The woman's hair was brushed out free, and was shoulder-length, curling under at the ends; one or two other women sitting in the room had the same hair-style, and in the lakeside film, Frances had seen it as the normal style for women.

The woman and Uxiaulia exchanged a few sentences in their speech, which seems rapid to our witnesses; and they laughed together. Uxiaulia turned to Frances and explained that they had been saying that *"looking at you people and the planet, it is like stepping back into one of our own history books: to us, you are living history"*. And they laughed again.

Uxiaulia went on to say: *"I must go now: someone is coming, and we have to move the ship. Every second we remain on the ground increases the risk of discovery. We shall not be going far; we shall set you down as soon as we can"*. And he and the woman hurried away.

Frances had understood that what was coming was a car; she said that somehow she knew that a car was unexpected in the place where they were standing on the ground, and took the ship people by surprise; afterwards she thought that, since the neighbourhood has much empty, unused ground with just an occasional rough track, seldom used, the ship had been set down in such a lonely spot where they were not likely to be disturbed: possibly a courting couple had driven their car, in search of a quiet place, along a track not normally used by traffic; or it could have been a police car on the prowl.

Most of the people in the room also left, until only some half a dozen remained. The 'big' man who had previously escorted Frances now came in, and indicated that she should follow him; they went along the corridor, still in the same direction. While walking along, Frances suddenly lost her balance and her feet left the deck; the man took hold of her and set her on her feet again, and they went on, the man walking behind Frances and keeping her steady with his right hand on her right shoulder and his left hand on her left hip; in this way they were able to make progress.

Frances could feel the ship rise up, and hear the deep humming note of the power generators below, low down in the hull; what caused her difficulty in balance was that a null-gravity field was in force throughout the ship, during the lift-off; later, in normal flight, gravity was restored. John had a similar experience, as we shall see, though it could not have been at the same time; it seems that the ship was lifted twice to avoid discovery.

They came to a room where Gloria and the children were waiting. Natasha said: "Oh, here's Auntie Frances", and the two women smiled; but nothing was said between them about their separate experiences. They settled down to wait for John; "Trust John to keep us hanging about", remarked his wife.

While waiting, they looked about them; the room was a fairly big one, of a curved shape: in the centre was a luminous cylinder which extended from floor to ceiling; it gave them a lot of difficulty trying to understand what it was. It did not seem to be quite like a solid thing; almost it gave them the feeling of a cylindrical beam of light, of uniform brightness from top to bottom, though as a whole it varied, being sometimes bright, sometimes dim. They could not see through it; and Frances was inclined to interpret it as a translucent cylinder illuminated from within. Later they understood its use, but not how it worked; it was a lift or elevator, but on a principle unknown on Earth.

There were some tables in the room, with chairs fixed by them; of course loose furniture is out of the question in a spaceship, as in an ordinary seagoing ship, and everything is bolted to the deck. A few people were sitting down. Two of the occupied tables had boards on them with buttons; the people were engaged in pressing the buttons in a rapid sequence, no doubt recording data of some kind, or making a calculation.

Set into one wall was a television-type screen, with three vertical bright lines on it, and short transverse lines of different lengths, set across the vertical lines: the whole pattern moved slowly down. This was evidently a repeater of a similar screen with identical patterns on it, which John was shown in the engine room, and was told it was concerned with the ship's altitude.

While Frances, Gloria, Natasha and Tanya are waiting for John to reappear, we will go back and pick up his story, from when he was taken for his medical examination.

CHAPTER SEVEN

John Examined

WHEN THE FAMILY separated after the speech of welcome on the balcony, John was taken along a short corridor to an examination room, similar in functional arrangement to the room Frances was taken to, but different in shape and detail. The officer who accompanied him was the same one who later showed John round the ship, and is almost certainly the one who made the welcoming speech.

He told John his name was Anouxia (pronounced 'Anno-Youksia'); he spelt his name out for John in the same way as Uxiaulia did for Frances, in a series of letters seen in the mind. Anouxia is clearly of responsible rank; and we would be inclined to call him 'Captain': but Uxiaulia told Frances that no one was in complete charge of the ship, which was run mainly by a computer.

Anouxia left John in the medical room, saying that he should wait there until someone came to do his medical examination; in fact John was left alone in the room for several minutes, and occupied his time, quite naturally, prowling and having a good look at his surroundings.

The shape of the room, as he describes it, is peculiar: the central feature is a black-upholstered 'dentist's chair' similar to that used by Frances; but the wall are as shown in the plan. There are two entrances to the room: one behind the chair, through which John entered, and another in front of a person sitting in the chair, not directly in front but somewhat to the right. This was not just a doorway, but a spout-shaped passage curving away to the right, with circular portholes, each about twelve inches across, on its left hand side.

Directly in front of the chair, but about twenty feet away from it, is a section of straight wall, without portholes, to which an instrument panel is attached. The panel extends from about head height to about table height; it projects a little from the wall, no doubt to allow space for components and circuitry. Laterally, the panel is five or six feet in length; directly in front of the panel is a desk or cabinet, of the same length as the panel and parallel to it; there is a space between the desk and the panel sufficient for a couple of people to stand and work conveniently; and the floor of this space is raised by a step to form a daïs or platform, so that a person standing behind the desk appears a little taller.

The desk has a sloping top, the lower side being towards the instrument panel and away from the chair; there are instruments on the sloping surface generally similar to those on the wall panel, which carries a variety of knobs, switches, controls, meters and little coloured domes that light up. The height of the desk is such that it forms a convenient working surface for a not very tall person who stands on the platform, facing into the room. There are no instruments on the vertical front of the desk, facing the chair.

Presently, two silver-clad persons came in; they were about five foot three inches in height, and of slight build. Although he did not immediately realise it, John later became aware that they were young women, very slim and small-waisted, with little breast shape as far as their silver uniforms showed. John said their skins were very smooth and fair; he thought they were quite young.

Only the faces could be seen: in common with many of the spaceship people of either sex, their silver uniforms continued over the head as a 'balaclava' helmet, coming up under the chin, and extended over the ears as a thin ear-shaped silver skin, fitting the living ear closely; so that their ears looked slightly larger than normal, but silvery. One assumes there must have been an opening, because these people heard perfectly well.

One of the two women went straight over to the desk and began operating the controls; she did not at any time speak to John, only smiled at him; he concluded that probably she spoke no English.

The other one dealt with John personally and worked around him; he saw her quite close, and says that she was

John Examined

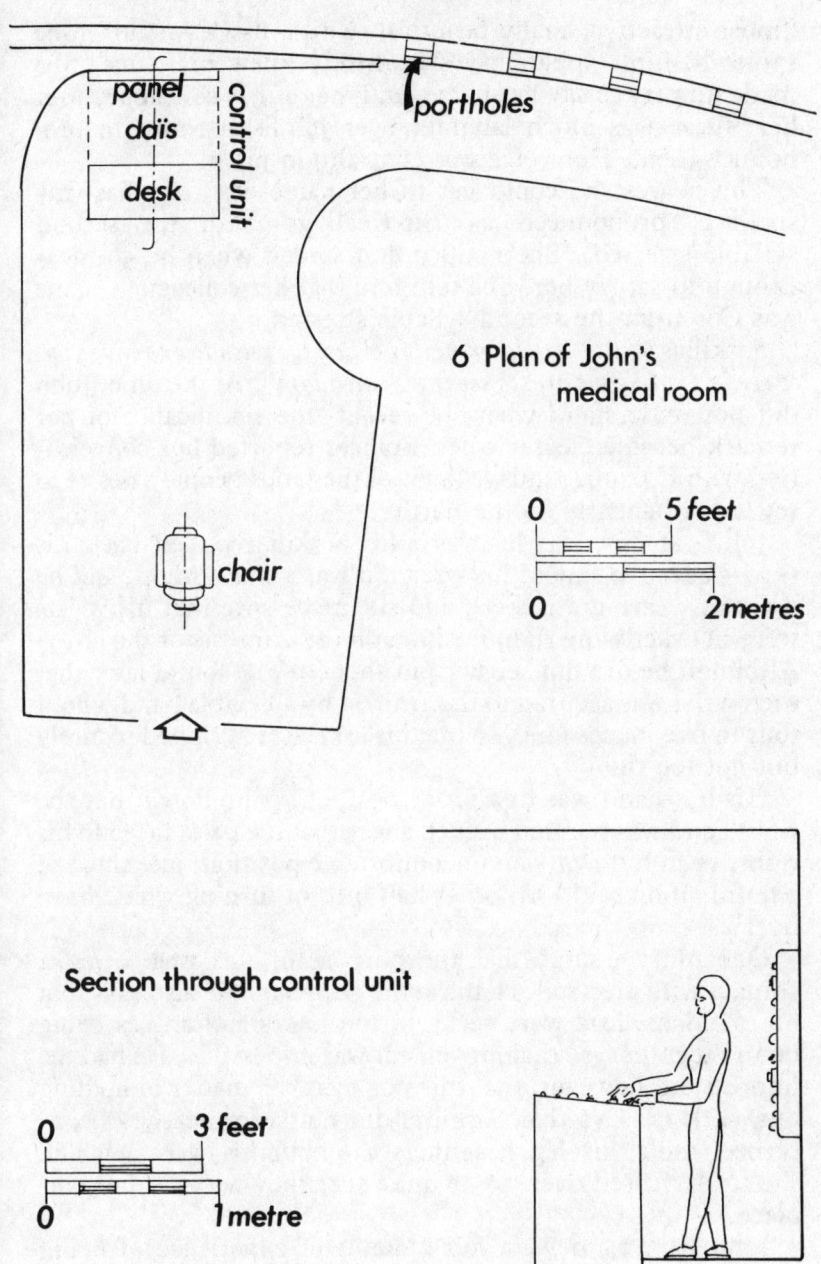

6 Plan of John's medical room

Section through control unit

"most attractive, really beautiful, with a flawless skin". She spoke to him; apparently she already knew his name. She made him try to say her name, and that of her colleague, after her; there was much laughter over John's attempts to pronounce them. Her voice was contralto in pitch.

The nearest he could get to her name was Serkilias (my spelling, pronounced as 'Sir-Kéelious' with the second syllable stressed). She nodded and smiled when he got near enough to satisfy her. She told him that her colleague's name was Cosentia, the second syllable stressed.

Serkilias said: "*Sit down in this chair; we wish to examine you, to see how alike we are: to see if we can adapt*". At the time, John did not understand what she meant; the significance of her remark became clearer when Frances reported her conversation with Uxiaulia, and we knew of the Janos people's desire to settle permanently on the Earth.

John sat down as he was told: Serkilias pulled back his shirt-sleeves to expose his wrists (it was a warm night, and he was not wearing a jacket); and she made sure that his wrists were in exactly the right position on the armrests of the chair. Although he did not see her put them on, he found later that each wrist was secured to the armrest by a flexible band, about four to five inches long, so that his arms were gripped securely but not too tightly.

His left hand was in a prone position, palm down; but the right hand was secured in such a way that the palm faced to his right, thumb down – an uncomfortable position, just short of painful. John said: "Another half inch of turning would have hurt".

One must assume that the purpose of this was to make contact with electrodes let into the armrests. He also says that he found his legs were secured, the knees and ankles being bound together, so that movement was impossible. He had the impression, however, that this was merely a matter of holding him still; he says that Serkilias did not seem nearly so concerned about his leg fastenings as about his wrists, which she really fussed over, to be quite sure they were in the right place.

John has, by now, a fair amount of experience of being hypnotised; although at the time of the spaceship adventure he

John Examined

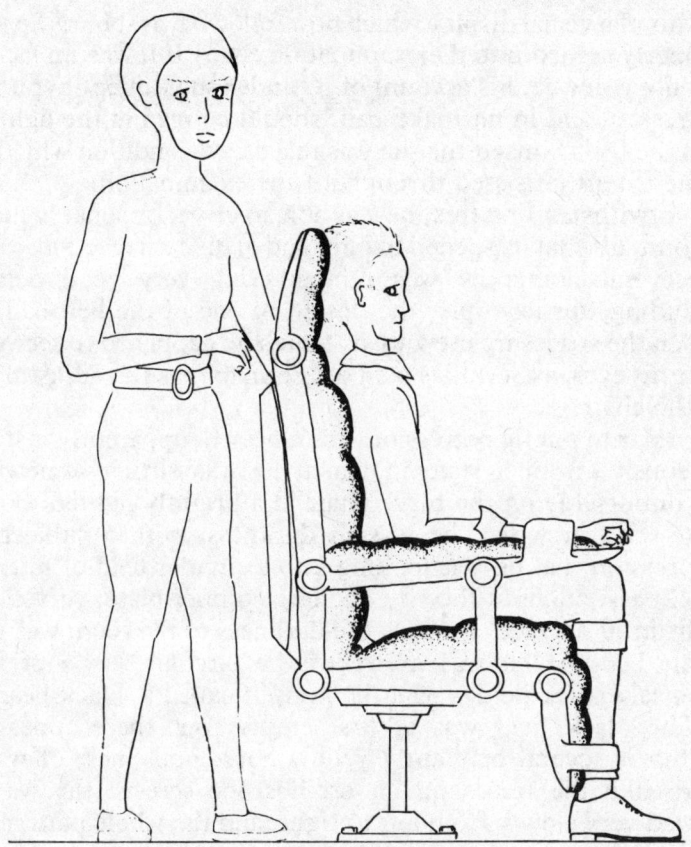

7 John on the examination chair, with Serkilias

had no such previous experience. He now feels that, throughout his medical examination, he was under hypnotic control: his memory is distinctly hazy concerning this part of his experience; and he says that there were times when he did not really know what was going on, though some details are clear. He had no will to resist the constraints and discomforts of the examination, some parts of which were not at all pleasant; he does not now feel that, without hypnotic control, he would have submitted so quietly.

Our hypnotist has suggested that John was hypnotised by

means of a visual display which now followed, as soon as he was properly settled into the examination chair. If it was, in fact, a hypnotic device, his account of it, under subsequent hypnotic regression and in normal recall, should be read in the light of this: he feels himself that he was in a dazed condition which to some extent persisted throughout the examination.

Notwithstanding this, he was able to give a reasonably lucid account of what happened to him, and of his own reactions; and at several points he was remembering very good detail, including, for example, the design of one of the belt-badges which the spaceship crew wear, which he happened to see very near his eyes, as Serkilias bent over him to make an adjustment to the chair.

Under hypnotic regression, John passed, apparently instantaneously, from a state in which he was sitting somewhat uncomfortably on the black chair in a brightly lighted room, to a state in which he was in darkness, with a subjective impression that he was looking into a circular field of intense blackness, of about the size of a large dinner plate, very close in front of his face. Perhaps the darkness of the room was not total; because he was aware of the circular shape of the intensely black field – what he himself called a 'black beam'.

The black field was at first empty; but there appeared within it several brilliantly yellow horizontal lines, of wave form like the traces on an oscilloscope screen; the waves flowed very slowly from left to right, and the whole pattern of wavy lines drifted down the screen, new lines appearing at the top to replace those that disappeared at the bottom. Mostly the individual lines were short, comprising only three or four wave crests; but later, some of these joined up, to form a longer line of many wave crests.

To anyone watching John as he sat in the hypnotist's consulting room, it was apparent that the wave pattern was causing him some distress and strain; for his eyes screwed up, like a person facing into a blindingly powerful light directed into his face and suddenly switched on. He said at the time that his eyes hurt.

After a minute or so of the wave patterns, the display changed abruptly. Two narrow ellipses, bright yellow lines on an intense black field, and crossing at right angles, pulsated

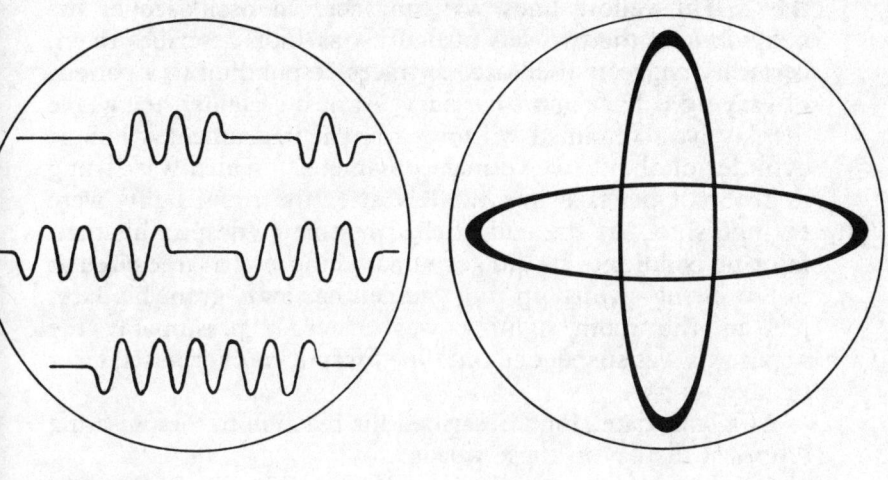

first pattern   second pattern

8 Oscilloscope patterns shown to John

rapidly in and out. The sequence is complicated: let us call the ellipses A and B, A being at right angles to B; then A and B both contract down to a very small circle, reaching minimum size at the same time. Circles A and B are concentric but of slightly different radii, so that each retains its separate identity. In the expansion phase which follows, what was originally the horizontally-extended ellipse A expands into the vertical extension; while B, which was formerly vertically extended, now re-expands into the horizontal plane. The two ellipses have thus exchanged positions. The frequency of the pulses of extension from ellipse to ellipse is 210 per minute. The two ellipses were indistinguishable from each other; but John was aware of their changing places at each extension.

This second display caused him more severe distress than the first; he says that, instead of merely a feeling that his eyes hurt, the second display gave him a burning sensation, right at the back of his eyes, which was decidedly unpleasant. He felt dizzy, as if he were about to faint; possibly he did lose consciousness for a time.

One must, I think, assume that the circular black field with the bright yellow lines was, in fact, an oscilloscope; the behaviour of the two sets of figures, as John describes them, certainly suggests oscilloscope traces responding to a pattern of varying voltage and frequency. Almost certainly, the whole display was contained within a circular instrument – a short cylinder of about twelve inches diameter – which was swung in front of his face immediately after the room lights were extinguished. At the end of this treatment, despite his half-fainting condition, he did get a final glimpse of a large circular object being swung up into the ceiling, away from his face, just as the room lighting was restored; presumably the apparatus was suspended on some form of counterpoised lever or lazy-tongs.

At a later date, John described his reaction to the pulsating elliptical display in these words:

"I had a very strange feeling when it went on to the second one. I had the feeling that you get when you feel you are going to faint; and you can't do anything about it – you can feel yourself going: you're floating, as it were. And I began to panic at that point; because I knew, or it felt that strongly, that if I gave up trying to fight it, I would pass out. It was just like the sense of fainting, when you try and fight off that feeling that you've got. But then it seemed to take me over; and I seemed to be floating."

He was allowed a short time of relaxation, in which to recover from the stress caused by the hypnotic display, if that was indeed its purpose, though it may have had other functions. No doubt some tranquillising effect was employed; because when questioned under hypnosis as to how he felt at this point, he said he felt "just peaceful", though there was still some residual soreness in his eyes.

I will continue with a somewhat condensed transcript of the talk between John and myself, which was the usual 'recap' session a few days after the hypnotic session in which he had so vividly re-experienced this part of his story:

FRANK: Well, you've experienced it yourself: when you're put under hypnosis, very often people describe themselves as having a floating sensation. Did you have that, at all, with Geoff?

JOHN: I have done, yes; but not as strong: this really felt quite intense, compared with –
FRANK: In other words, you had no sense of up or down, or gravity?
JOHN: No; I just seemed to be completely suspended, as it were. I could still feel my body pressed into the chair; and when the lights finished, one of the people came right up close to me – that's when I got a good view of the badge on her belt: and then the whole weight of me, of my back in the chair, I could feel it change to as if I was laying down; I could feel the weight of my legs pressing downwards.
FRANK: One moment you were weightless, floating; is that right?
JOHN: That's right; and then I had the feeling that I was being pressed down, and the chair had tilted.
FRANK: Your chair had tilted? – we didn't have the tilt before.
JOHN: Yes; the chair definitely tilted.
FRANK: Just like a dentist's chair tilts back?
JOHN: Tilted back, yes.
FRANK: Did it open out into a bed, almost?
JOHN: Yes: this is what I meant when I said I could feel my legs pressing back; because in the first instance my legs were bent and I was sat; and as the chair tilted back I could feel my legs straighten out, and they were being pressed down as well.
FRANK: So the change from the floating feeling to the feeling of being pressed down came just as the chair tilted?
JOHN: Yes. And then after a while – one of them was standing by me; I don't know what she was doing, because I couldn't feel anything: I felt numb all over, apart from this feeling of being pressed – after a while, I could feel my back pressed into something; and then that changed to the same feeling, but pressed all the way down my side, as if I had been turned over. [On another occasion, John gave me the impression that he remained for some time on his back, while instrumental readings were recorded; and that he was then turned on to his side, while more readings were taken.]
FRANK: Which side?
JOHN: I was on my right side. Then, after a while, I was

turned back to as I was, on my back still laying down; and then I felt my legs go down as I came up into a sitting position again.

FRANK: Of course Frances describes this very strongly: this being pressed down into the chair. She said as if her weight had doubled. Her chair is very similar to yours; so I would expect it to work in the same way. Her drawing of the chair is very like yours; in fact, hers is more elaborate, very detailed in fact; she draws the detail of the attachment to the deck with octagonal bolt-heads on it – almost an engineering drawing; she gives measurements. And she says she is quite positive that it really was like that. [Readers who are familiar with the expression 'nuts and bolts ufology' may like to make a note of this.]

JOHN: Well, I know I was laying down; because it was at the point where I felt the chair go back, that I realised where the bright yellow light was coming from: it was straight above me; it was just like the dentist type circular light overhead which is a white sort of colour, but this was a bright yellow – just like the sun. Sometimes it was a yellowy colour, and sometimes it was a bluey colour.

FRANK: Can you tell: was it the only light in the room?

JOHN: No, I don't think it was; where the people were standing by the instruments seemed to be more of a white light. So I think there was more lights in the room.

FRANK: So it was just a spotlight over the chair?

JOHN: Yes.

John said that he had at one point a good close look at the badge on the belt worn by Serkilias. It was circular, slightly wider than the belt, about three inches diameter: the design consisted of a white disc bearing a stylised representation of a 'flying saucer' ship such as the one they were in, as seen in side view; from the centre of the under side, lines representing the limits of a slightly divergent beam extended to the limit of the disc, in a downward direction. The design was delineated in black lines. Surrounding the white disc was a narrow black annular zone, perhaps half an inch wide; it bore a number of short straight silver lines, set at odd angles, forming a kind of pattern all the way round. John said it was not regular enough

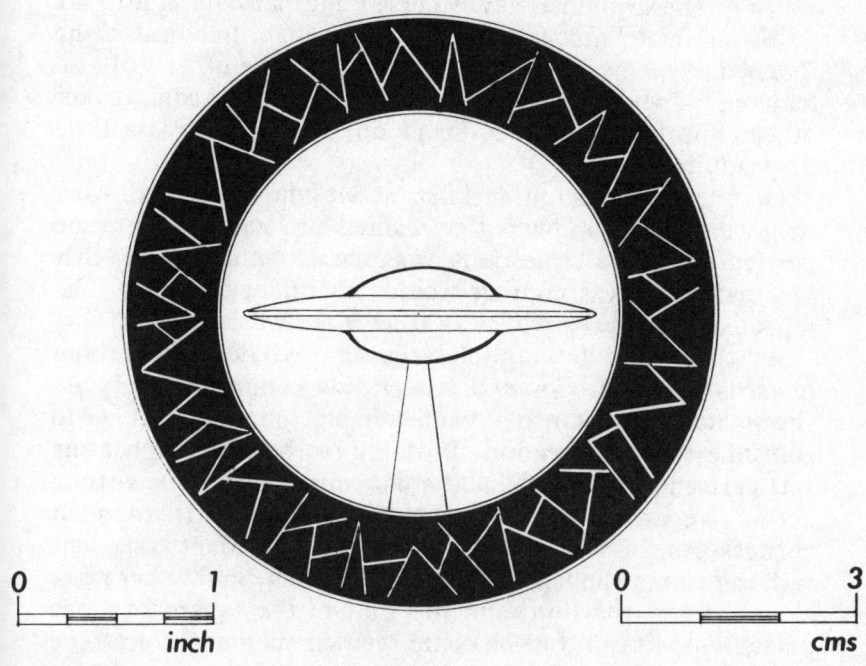

Badge worn on the belt by members of the spaceship crew

ALTHOUGH THE CENTRAL SPACESHIP DEVICE IS CORRECT, THE PERIPHERAL INSCRIPTION IS TO BE TAKEN ONLY AS A GENERAL IMPRESSION OF STYLE, AND NOT AS AN EXACT COPY

to be a pattern; he had a feeling that it was writing, that it meant something.

He was still in a somewhat dazed condition when Serkilias came over to tell him his examination was complete. She had to repeat it several times before she could get through to him; finally she smiled as John's response showed that he understood what she was saying; she said again: "*Your examination is complete, John*".

Serkilias gave him to understand that they had to wait for someone who would come and fetch him. While they were waiting, she explained about herself and her colleague: *"We do the medicals; that is our job"*. As she said 'medicals', she touched a yellow band on her shoulder – the other girl also had one – perhaps indicating that this was a badge of her office. She helped him to stand up; at first he was a little unsteady on his feet.

He tried to question Serkilias about who they were, what they were doing, and why they wanted him examined: but she parried adroitly all questions of a general nature, saying that the man who was coming would tell him everything. She repeated: *"The examination is to see if we can adapt"*.

As they were talking, Serkilias moved down the room towards the desk; Cosentia was already behind the desk. As she walked the distance of ten feet or so, John followed her to continue the conversation. In doing so, he noticed that she was carrying in her left hand a flat square silvery-metal box, about five inches square by one inch thick, with rounded corners. As she was saying "to see if we can adapt", she was walking round the left hand end of the desk, so that her right side was towards him, and his view of the square box was partially obscured; but he could see that she put it into some kind of slot in the face of the instrument panel, on a level with the desk top, but in the wall panel, not in the desk itself. He was, by now, on the room-ward side of the desk.

John had an uncomfortable feeling; he said "what have you done?" or words to that effect, indicating the square box.

Serkilias replied: *"Samples, They are going to be analysed. We have taken blood samples"*.

Now John, as it happens, has a particular dislike of having blood samples taken; even the prospect of it makes him go hot and cold, and feel "horrible", he says. He had this feeling very strongly now. He says that the two girls realised he was alarmed and upset, and they were smiling and laughing in a friendly, reassuring way to let him know it was O.K., and not to worry.

Presently, the man they were expecting did come in: it was Anouxia, who had brought him to the medical room. John now noticed that, instead of a belt badge. Anouxia had on his chest an absolutely plain white disc, of about five and a half inches diameter, with no visible markings on it. (Uxiaulia, who talked to Frances, had a similar disc.) He was a man of about John's own height and build – John is six feet tall, and slim – and he wore no helmet; his hair was fair and cut very short. He had blue eyes, like all the ship people.

Anouxia first talked with the two medical technicians in their own language; it sounded rapid to John, as foreigners' talk is apt to do. They kept glancing in his direction, and seemed to be discussing his case.

Finally, Anouxia came over to John, who had moved aside. He said he would show him over the ship, and answer his questions.

John took his leave of the two women, who smiled goodbye; and he followed Anouxia along a corridor, which was rather dark; there was just enough light to see by. He noticed that the corridor curved slowly to the left as they walked: in the right hand wall, from time to time, they passed a circular porthole of thick glass; John could see only darkness through the portholes.

Presently the man stopped in front of him, in a completely dark space; John almost bumped into him, but he could just make out a faint silvery gleam on his shoulders.

He was just wondering why the man had stopped, when he felt himself floating slowly downwards: it was like being in a lift, he said, but much smoother, and silent. Although he could not see anything around him, he had a sense that the space was circular; perhaps there was a very little light, because he could just make out his companion.

They floated downwards some way, then came smoothly to a stop. Anouxia led the way, as before; they left the elevator, or whatever it was, not by the way they had come in, but to the left: John was quick to realise that the corridor must have followed the curve of the hull; so that this left turn would take him towards the centre of the ship.

They came out through a doorway, on to a balcony overlooking a huge circular room (John estimated its diameter as 150 feet) with many pillars or columns supporting the ceiling. It took him a little while to realise that this was the same balcony on which he and his family had been welcomed by Anouxia and the other officers of the ship: that the vast circular room was the same that they had entered from below, through a big square hatchway, from the airlock entrance under the ship. Somehow it looked different, coming to it like this, from the other direction.

John says that he asked where the others were, meaning his family; Anouxia replied: *"Do not be alarmed; you will see them soon: they are in their medical, and are being shown other parts of our ship"*.

Anouxia seemed preoccupied; and John had time to look about him. Just in front of him was a handrail; the balcony stood, he thought, some ten feet or so above the circular deck below. The handrail was shaped just to fit a human hand comfortably; the space below it was filled in with metal panels. To his right, the balcony ended, the handrail being turned through a right angle until it met the wall behind. To his left, the balcony, with its handrail, continued in a downward slope to the main deck; and he remembered the moving ramp up which they had glided, going up to the balcony when they first arrived.

Right opposite he could see on the far side of the room a similar balcony; he could see two arched doorways below it, and doors through the wall behind the balcony, as there were on his side. There was a ramp, also, on the other side; but whereas the ramp on his own side went down to his left, the ramp on the far side also went down to the left as John saw it; so that to a person standing on the opposite balcony, the ramp would be to his right.

The balcony-ramp system formed two opposing segments of a circle – not the circle formed by the extreme limits of the great room, but another circle, concentric with it, and somewhat smaller. Immediately behind each balcony-ramp segment, a complete vertical wall extended from floor to ceiling, so that one could not see into the space beyond; but in the intervening larger segments, where there was no wall, ramp or balcony, one could see right through to the distant outer wall of the circular room.

John noticed that, whereas within the balcony-ramp circle the floor was flat, beyond it the floor rose up in a smooth curve, so that its outer edge, where it met the wall, was several feet higher. He had already seen the bowl-shaped underside of the ship from below, while they were still on the ground; and he guessed that the reason why the outer part of the floor curved up was because in that part, it had to follow the curve of the outer hull of the ship.

There must, he realised, be another space, at least around the centre, between the flat part of the floor and the outer hull; and when we came to make drawings we understood that, in the middle, this lower space must accommodate the large airlock; there is, in fact, deck space around the airlock which John did eventually get into.

Near the middle of the circular main deck, his eye was caught by a large oblong white object: this particular recall first came through hypnotic regression, and – characteristically of such recalls – it was vague to start with, then sharpened up into detail as he concentrated on it. The white object came into focus; and he recognised the car he had been driving when they were intercepted by the flying saucer.

There was a loud buzzing noise: and Anouxia leaned over the corner of the balcony, as if he were looking down across the main deck. John noticed, however, that the movement had the effect of making him face into a small rectangular box, mounted on a short metallic column upon the corner of the handrail. The box was covered with a fine metallic mesh on the side Anouxia was facing; and there was a control knob to one side. It suggested a small loud speaker; but when Anouxia began speaking into it, and his voice, enormously amplified, came back from hidden speakers elsewhere, John realised that it was a microphone, feeding into a public-address amplifier.

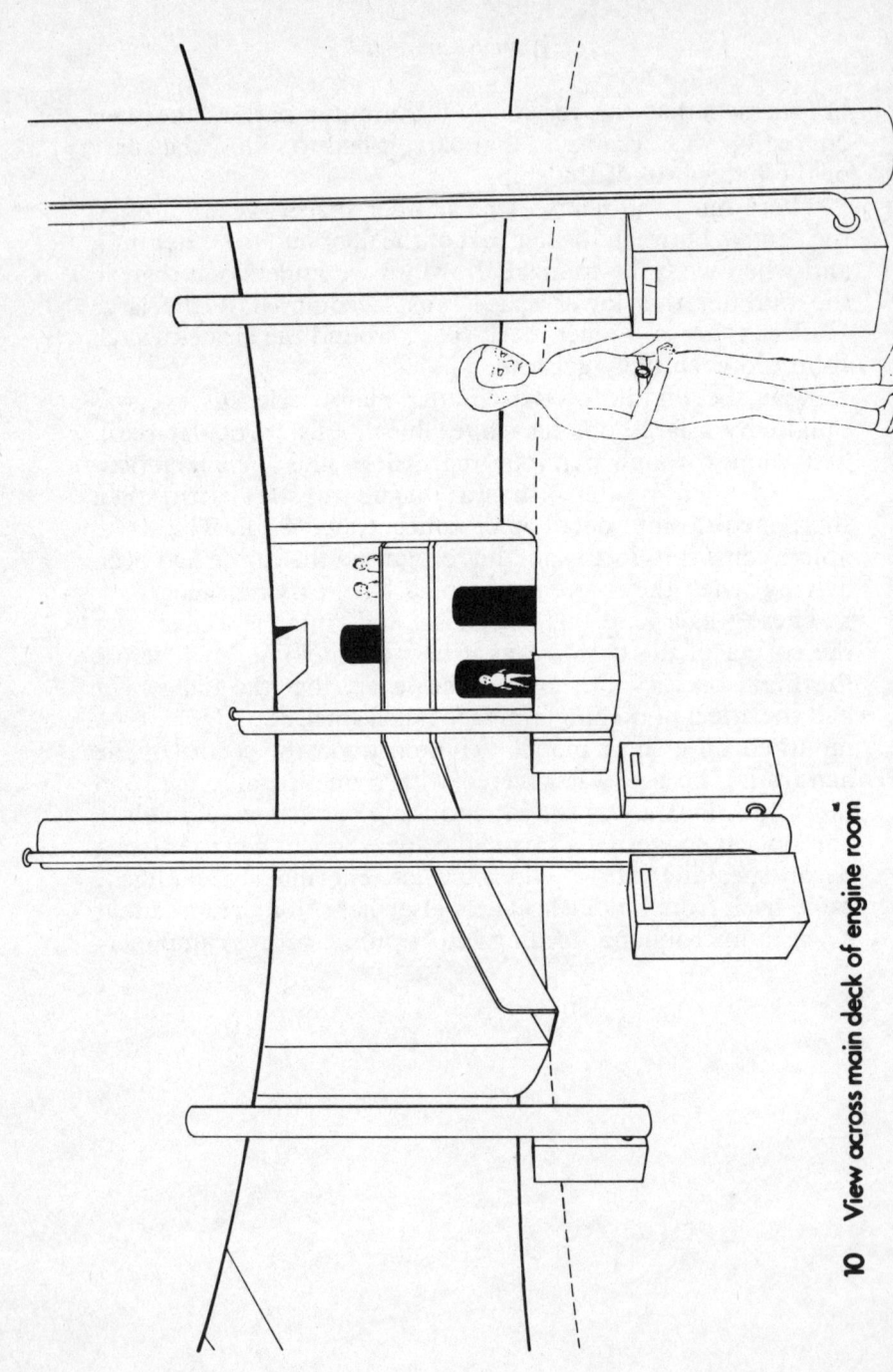

10 View across main deck of engine room

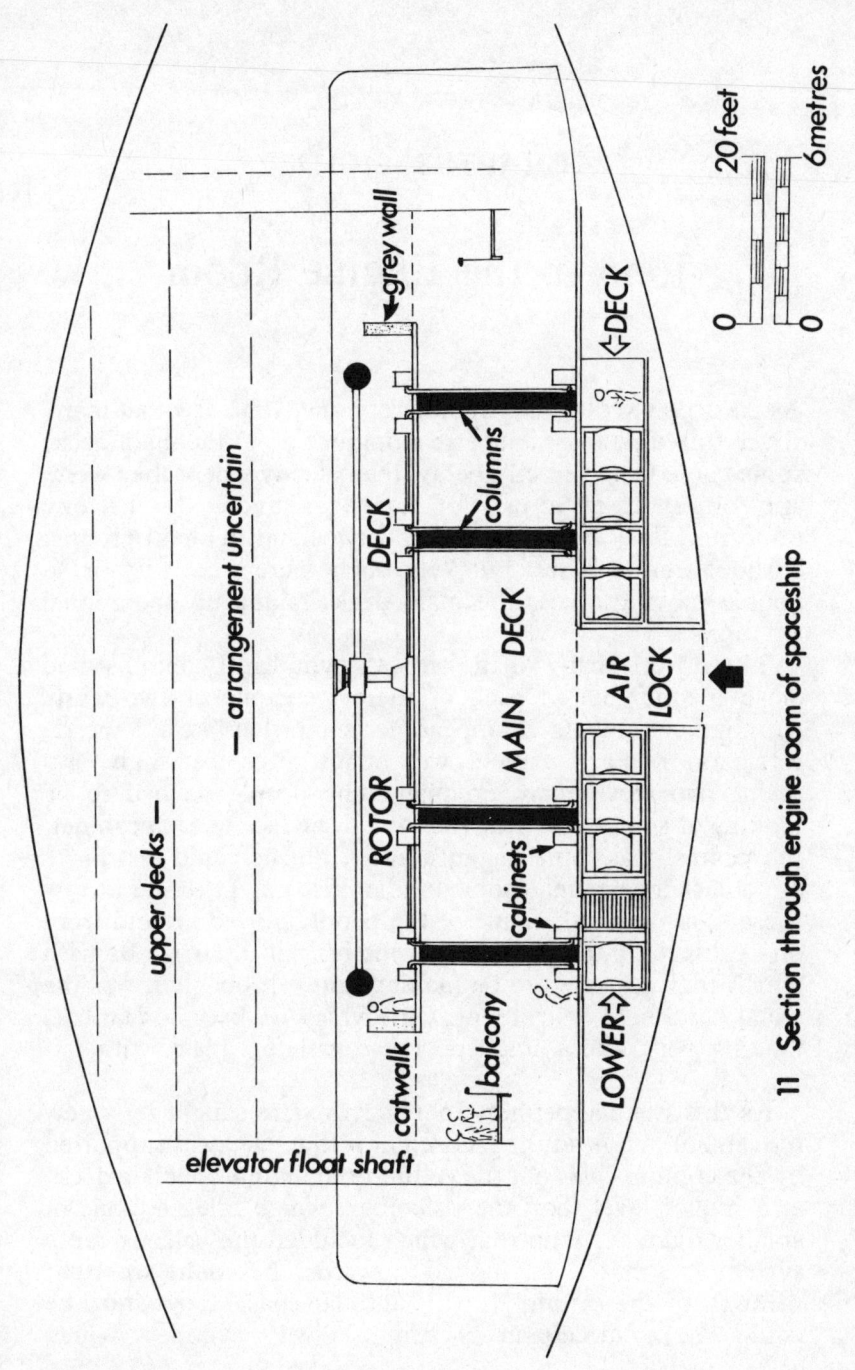

11 Section through engine room of spaceship

CHAPTER EIGHT

John in the Engine Room

As ANOUXIA SPOKE into the microphone, John saw that many silver-suited people came streaming out on to the main deck; some came from directly below the balcony where they were, and others from a pair of doorways under the balcony opposite. They came in ones and twos and in small groups, without undue haste; but very soon there were fifty or so people moving across the main deck, taking up operational positions.

There were many of the smooth cylindrical columns; and close by the foot of each of them were one or two white rectangular cabinets, in appearance somewhat like a domestic refrigerator. Each cabinet was about three feet high, and about two feet square on plan. The people seemed to be looking at something near the top of one face of each cabinet; the nearest was some distance away, but he could just make out a darker rectangle, some kind of instrument. Later he saw these close at hand. Some of the people moved around from one cabinet to another, with something in their hands with which they appeared to be making notes; probably it was the usual hand-held instrument, with which we became familiar, looking very like a small pocket calculator, black with red buttons.

As this was happening, John's eyes were caught by a new movement overhead; he realised now that what was supported by the columns was not the ceiling, but an intermediate deck, at a higher level than the balconies, which filled a circle of smaller diameter than that which included the balcony-ramp system, so that if he looked upwards, he could see right through to the ceiling. In this annular space above him, he could see a succession of bright silvery cylinders which travelled round in a circular path, at first slowly, but with rapidly increasing speed.

Each cylinder was attached to a radial arm which came from the direction of the unseen centre of the ship. Soon he realised that there was just one long beam, pivoted centrally somewhere out of his sight, with a cylinder at either end. Each cylinder was shaped, if one can imagine such a thing, like a double-ended bullet; the middle part was cylindrical, but each end tapered off into a streamlined paraboloidal 'nose'.

Anouxia now turned to John, and, seeing him looking at the bright cylinders on their rotor beam – they were now turning much faster – said: "*If we turn fast enough, there is no gravity to hold us down to the Earth*". He went on to explain that he had been telling the people to get ready to raise the ship off the ground, because someone was coming, and they had to move to avoid discovery. Under hypnosis, John said: "He has told me that someone is coming; not to be alarmed: someone is coming; they must move on before they are seen. When they stay on the ground, it is danger to them; they must move. They are frightened of being captured; they want to go."

A humming sound, quiet at first, coming from below the main deck, increased steadily in pitch and volume, until it reached a fairly noisy maximum; but the giant rotor continued to increase its speed until the separate repetitions of the passing cylinders could no longer be distinguished, and the whole visible part of the rotor became a uniform, gleaming silvery disc. John says the engines made a lot of noise, and he could feel the deck vibrating under his feet.

At this point, he had a curious experience: he became weightless, and lost his balance, falling helplessly sideways, but not falling to the deck. He said under hypnosis: "It's like falling into water". Anouxia burst out laughing at this – just as a sailor may be amused at the efforts of a landsman to keep his feet in a rough sea; but he grabbed hold of John, and set him on his feet again. No doubt the joke was an old and familiar one.

We may recall that Frances had a similar experience while walking along a corridor after her talk with Uxiaulia; but an analysis of the timing of the whole visit makes it doubtful whether these can have referred to the same lifting of the ship.

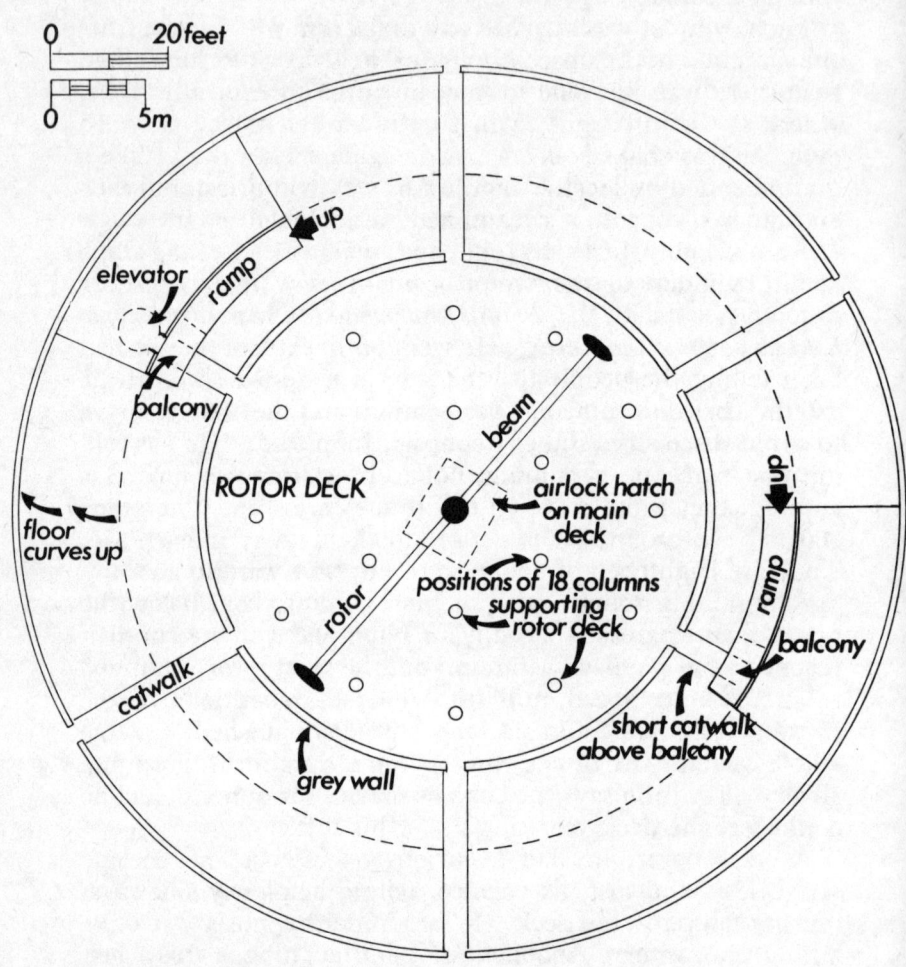

12 Plan of upper levels of engine room: cabinets and pipework on the rotor deck are not shown

John in the Engine Room

Frances, as well as her medical, had spent a long period with Uxiaulia, talking and watching films, before he was called away to move the ship; whereas John had just finished his medical, and his two very prolonged visits, one to the engine room and one to the navigating bridge, were yet to come when he felt the ship rise. It seems that the ship moved twice to avoid discovery.

He could now feel the entire ship stir and rise, tilting through an angle as it did so. Anouxia pointed to the thick white soles on his own black shoes, and demonstrated to him that they enabled him to walk securely on the deck; they seemed to cling to the deck, but could easily be lifted clear when he walked.

Under hypnosis, John said: "I can hear the engines, and I felt the floor vibrating – I felt we were going up very fast – falling over backwards – floating back – can't stop – he has stopped me there – grabbed hold of me – pulled me back".

GEOFF: What prevents him from falling?
JOHN: He is pointing to his shoes; he is pointing as if to show the thick white soles. It's like foam rubber. He's pointing at them, and putting his foot back on the floor. They seem to stick. He's laughing. Floor tipped, and left side came up; I felt like in a plane, when it banks.
GEOFF: How long did that feeling last for?
JOHN: When I was falling over – it only lasted for about – ten seconds. The floor tipped, not for long, few seconds; then it levelled out . . . Going back down; landing again.
GEOFF: How do you know?
JOHN: The engine noise; it's dropping. I can hear the rotor going slower. The engines go slower; and it seemed to bounce – went down, popped up slightly . . . Think we landed; rotor's going slower still.

Anouxia told John about the big circular room: *"This is where we make power"*. He said something about electro-magnetism; he said they used *"very, very high voltage"*. John understood him to say that the humming sound he heard coming from below was a starting device; once the rotor was spinning fast enough, it "took over".

When the crisis was over, and most of the crew had dispersed to their quarters, Anouxia laughed and said to him: *"You were lucky to be here and see our engine working".*

Anouxia led John down the ramp, on to the main circular deck. They walked across to look at one of the white cabinets, at the base of a column. He now noticed that a white pipe, of about five inches diameter, emerged from the back of each cabinet, low down, and ran up alongside the nearby column, penetrating the perforated rotor deck overhead. Later he was to find that the 'pipe' – which one must suppose to be a heavily-insulated cable – connected with a similar cabinet on the rotor deck, directly above.

Anouxia opened the hinged side of the cabinet to show him what was inside the casing, but it was not very informative. He could see only a large black rectangular block, with no detail on it, which almost filled the casing, leaving only a small air gap. One may suggest that the black block may contain a condenser, also heavily insulated. Frances was told that 'static electricity' was important in the power system of the spaceship.

Near the top of the front of the casing was a recessed instrument shaped like a letter-box, with a linear scale on it; a needle was registering on the scale, about three-quarters of the way across from the left. To John it suggested a voltmeter, and it may well have been just that; Anouxia pointed to the meter, with its needle well over to the right, and said: *"High power".* He went on: *"When we lift the ship, much power is drained away, and we have to watch that too much is not used".*

Nevertheless, it seemed odd that so many people should be needed to take separate readings, when the whole thing could easily be handled by a computer. He had a feeling that Anouxia was trying to explain why they needed so many people, but his memory of this is not clear. What we thought ourselves – and it may have been suggested to John, but we cannot be sure of this – is that, because so many people have to be carried in the spaceships, they were really making jobs for people, especially younger men and women who might become bored and restless with spaceship life, to give them something to do, and some experience of an actual planet-landing, which must have been exciting for them, after being so long in the ships.

They walked round the huge circular deck: at one place, set into the outer wall, was a pair of television-type screens; below each screen was a broad shelf with some instruments and controls on it, and a seat of the usual one-legged pattern was fixed before each of the two screens. One of the screens was blank; but the other was in action, and a silver-suited technician sat on the black saddle-seat, watching the display on the screen, and periodically making adjustments to the controls. The luminous display consisted of three vertical lines, extending from top to bottom of the screen; placed across these lines were a number of short horizontal bars of varying length, one set of bars to each vertical line. The whole display was drifting very slowly downwards. John was told by Anouxia that this instrument had to do with the ship's altitude.

Anouxia next took him to a place in the main deck where there was a flight of steps down to a deck below; it was protected by handrails. It is noteworthy that this is the only occasion on which steps or stairs are referred to; I questioned John about this, but he was quite sure that this was, in fact, a fixed flight of steps, rather steep.

The lower deck to which the stairs – perhaps we should say companion-way – led down was much less extensive than the main deck, which is what we should expect. It was also of low height; the deck-head was about eight foot six inches high.

Towards the centre of the ship, a bulkhead obstructed his view; this must have been the outer casing of the air-lock. Elsewhere, passages led off in different directions; on either side of each passage was a series of huge shapes, each consisting of a big square frame, filling the space available vertically, and about six feet in width.

Extending horizontally across the middle of each frame was an elliptical cylinder – if geometricians will permit the expression – about five feet long, the elliptical cross-section being about four foot six inches in major (vertical) axis, and about two foot six inches in minor (horizontal) axis.

The whole thing was cased in with an off-white smooth material, possibly a plastic as we understand the term; so the

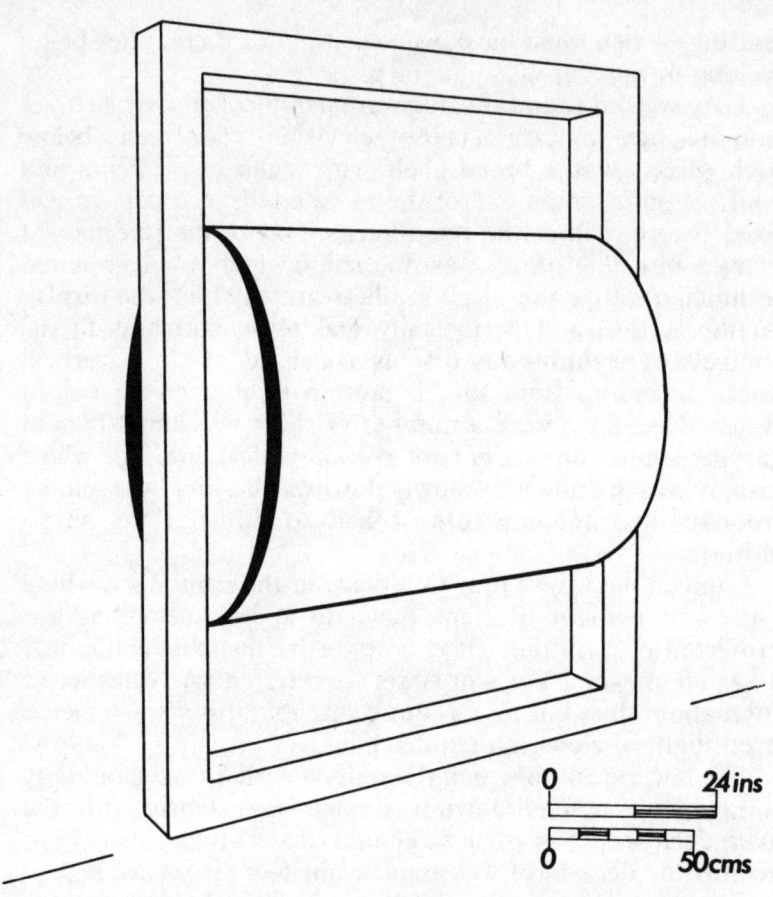

13 A transformer? unit from the lower deck of the engine room

internal details could not be seen: but the shape and arrangement strongly suggest a transformer. One must assume the concealed presence of a horizontal member at middle height across the frame, forming a core for the windings; the elliptical shape no doubt allows for better use of the available space, and easier access for maintenance.

Again, although the casing did not allow it to be seen, it is reasonable to assume that the whole frame, including the unseen cross-member, is laminated in the usual way to discourage eddy-currents; and that it forms, as a whole, a core to concentrate the magnetic lines of force.

Although he was shown only a part of the lower deck, John understood from Anouxia that it was all the same; and that the deck space was filled with these power units, allowing only the necessary access ways for service. Even if aluminium wire were used for the windings, the total weight would be considerable; it would have the effect of concentrating a large part of the mass of the spaceship into the lowest part of the hull.

They returned to the main deck; but John's tour of the ship's power-complex was not yet finished.

Earlier he had noticed that the numerous columns supported a circular deck, less in diameter than the circle which included the balconies, the deck surface being made up of long panels of perforated metal, about four feet wide, arranged radially in a six-rayed pattern. The panels were cut to shape towards the centre, so that they all lay in one plane; he presumed there was a structural framework to hold the whole thing together, but could not make it out.

Because of the perforations, he could to some extent see through the upper deck from below; he could make out some white cabinets similar to the ones on the main deck, but with an important difference – a great many pipes, mostly thin but some of them very thick, extended from cabinet to cabinet, forming a spaghetti-like maze, so dense that he had been unable to see right through them to the final ceiling above. The thick pipes were similar to those he had seen running vertically, close by the columns, joining the cabinets on the main deck to those on the rotor deck above; but the numerous thin pipes, about an inch in diameter, ran from one cabinet to another on the rotor deck only – there were none on the main deck.

When he first looked up through the outer part of the perforated rotor deck, before the cylinders of the rotor began to sweep around, there had been some half-dozen or so technicians working up there; but when the rotor began to turn, he had noticed that they left the area in something of a hurry.

John had also seen that the perimeter of the upper deck, what we came to think of as the rotor deck, was enclosed by a thick circular wall of a dull grey colour – this is noteworthy as one of the few exceptions to the ubiquitous white or silvery surfaces of almost everything in the spaceship. At six points, equally spaced round the circle, the grey wall was interrupted by a gap of about the width of a normal doorway, which gave access to the rotor deck itself: extending radially outward from each of the six gaps was a catwalk of the same perforated metal – just a single four-foot-wide strip, bridging the space between the wall-gap and a doorway in the outer wall. To be more precise, although he did not realise it at the time, a design analysis shows that four of the six catwalks extended right out to the outer wall of the engine room, but two of them, forming an opposite pair, ran across to meet the somewhat nearer wall segments behind the balconies.

Anouxia now took John back up the moving ramp to the balcony again; they passed through a doorway in the back of the balcony, and he found himself back in the dark lift. This time they floated upwards: not very far; and they came out on the same wall that was behind the balcony, but higher up, still within the great circular room. Now they were on a level with the rotor deck, and Anouxia walked out without hesitation along the catwalk, although it had no handrail, and there was nothing to stop one from falling to the main deck, some twenty feet below.

John followed Anouxia along the catwalk to the gap in the grey wall, which they passed through on to the rotor deck itself, with its white cabinets and the maze of white piping connecting them. The piping was low down near the perforated deck; and now he realised that the cabinets were lower than those on the main deck, being only about two feet high, though their other measurements were similar.

The giant rotor was now still; and he was able to examine it closely. In fact, Anouxia told him, no one was allowed to be on this deck while the rotor was turning: it was too dangerous.

The beam which carried the bright cylinders on its outer ends was well above John's head; its lower face was nearly eight feet clear above the deck on which he stood. The beam was sixteen inches wide; its edges were rounded, and the

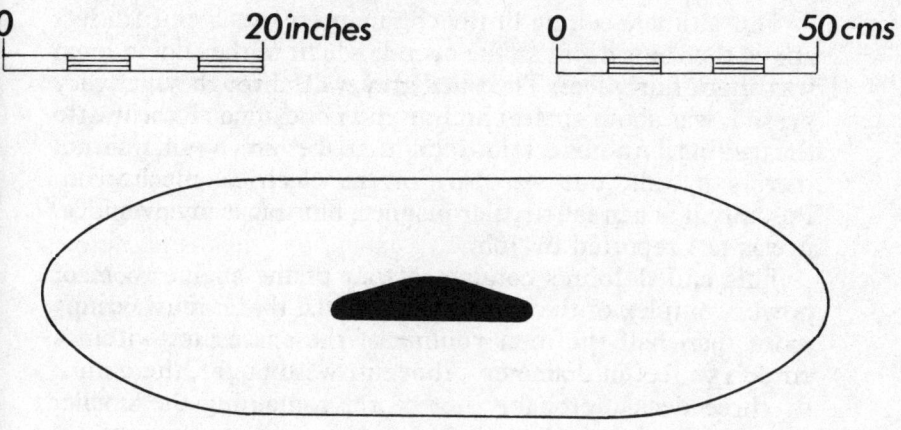

14 Profile of a rotor cylinder with section of attached rotor arm

upper surface sloped gently up towards the middle from each side. This sectional shape would give some stiffness without impairing the transverse streamlining, so important when the rotor is turning at really high speeds; it would, I think, also give a slight aerodynamic lift to the rotating rotor beam, which was about 36 feet in overall length.

At each end, the beam was inserted laterally into the centre of the cylinder; each cylinder was about five feet long, and about 18 inches in diameter in the middle straight-sided portion. This cylindrical portion was about 27 inches long; beyond that, the shape smoothly tapered off like a blunt-nosed bullet, with well-rounded ends, not pointed.

The maze of pipework made it difficult to approach the centre of the rotor deck; but John could see plainly the central pivot about which the rotor beam turned: it consisted of a vertical shaft, about 18 to 20 inches in thickness, extending from floor to ceiling, being secured top and bottom in a massive bearing. A comparably massive collar, mounted on the shaft near its upper bearing, provided secure anchorage for the beam. My calculations suggest that this collar occupies the geometrical centre of the ship.

The ultimate ceiling of the engine room was about ten feet above the rotor deck; so the overall height of the engine room was about thirty feet. The thick grey wall, through which they passed, was about six feet high; it did not extend all the way to the ceiling. Anouxia told John that the grey wall was not merely a wall, but was part of the electrical mechanism. Possibly it is a great circular magnet; but this is my own idea: it was not reported by John.

This ended John's conducted tour of the engine room or power complex of the ship; we calculated that it must occupy more than half the total volume of the spaceship, within a circle 150 feet in diameter. Above it, we thought, there must be three decks, probably not more, containing the smaller rooms, some of which our witnesses have spent some time in, though we did not find it possible to assign precise locations to them; there must be living space for upwards of fifty people, as well as all the remaining stores and technical facilities. One of these latter, the navigation room or 'bridge' as we might call it, John was next to visit.

CHAPTER NINE

Pictures in the Navigating Screen

To REACH THE navigation centre, Anouxia took John once more to the elevator; this time they floated upwards for some distance, then emerged into a large room set about with banks of complex instruments on the walls and on control desks around the periphery of the room.

The centre of the room was occupied by a big horseshoe-shaped desk, about three feet high and about two feet six inches wide; the inside diameter of the horseshoe was about nine feet. A chair was fixed in the centre of the concave side, and another opposite to it; other chairs were provided for other parts of the great table. Some of these other chairs were occupied by silver-clad persons.

Anouxia went to the chair in the centre of the concave side, and indicated that John should sit opposite to him. The desk surface was divided into two equal contrasting zones: to Anouxia's right hand, the entire surface of the desk was crammed with instruments, knobs, switches and little coloured lights; John said it was so crowded, you could hardly have got another one in: to his left hand, the desk surface was black, smooth, bare and featureless.

Indicating the desk, and the room as a whole, Anouxia said: *"This is for navigation"*. (I note with some satisfaction that the Janos people do not use made-up 'space' words, such as 'astrogation', beloved of some science fiction writers; when speaking English, and presumably translating from their own idiom, their terminology is derived mainly from the language of the sea. The very craft in which they live and travel is, to them, a 'ship' – much less frequently, a 'spaceship'. When Anouxia first welcomed our family aboard, he said: *"Welcome to our ship"*. Nothing more clearly indicates the original maritime nature of the Janos people; we had many occasions to note their passion for anything to do with water and with craft that sail on water. A flying saucer is, to its makers and crew, a ship, not an aeroplane.)

John, unfortunately, has never quite recovered his full, detailed memory of sound, including speech, during this part of the incident; though he did eventually, after some hard work and persistence on the part of the hypnotist, recover enough speech-memory to get the main substance of the story, with some important verbatim passages. This may be partly why one tends to get the impression that Anouxia's English speech is less fluent and less idiomatic than that of Uxiaulia; though it would be reasonable to expect some individual variation in aptitude for languages.

John therefore missed quite a lot of the sound-track, as it were, of this first section, in which Anouxia tried to explain to him the operation of the complex control system. Anouxia kept on pressing buttons and turning knobs; and little coloured lights would light up: but without the speech, John does not really know what it was all about. It seems, moreover, that he did not understand at the time, that is not merely that he does not remember understanding; for there were other passages in which he knows that he heard and understood at the time, but does not yet remember what was said. John does remember that Anouxia, on this occasion, realised that he was not taking it in, and repeated the whole demonstration.

It should be remembered that John, while an intelligent man and skilled in his occupation, does not have the specialised training which might have made some sense of the button-pressing; without it, and without most of the speech-memory of this part of the incident, a full understanding is hardly to be expected. I suspect that his difficulty in remembering speech in this highly technical section is caused by his lack of understanding.

It may be, too, that scientific principles were involved which would be unfamiliar, even to a scientist; though in general, we find that we are able to understand the basic science behind what the Janos people are able to do: where we cannot follow is in their technology; how they do it. When it

Pictures in the Navigating Screen 109

comes to their apparently complete control of gravitation and inertial mass, we cannot even follow the basic science.

John does remember, at this point, that he asked Anouxia: "Where do you come from?" – and Anouxia answered: "*I will show you*".

They moved over to the blank half of the navigating table, and at once a large rectangular area within the smooth table-top lit up as a screen; there was room for a second screen beyond it, John says. At first he could see nothing but a uniform blue background, which gradually darkened as various planets began to appear on it.

At first he saw what he called under hypnosis "a large round like a ball, with curved ridges on it. Like mountain ranges on it. About six inches across. Brown with a golden tinge to it". (By "six inches across" he means the actual size of the image, on a screen of about 42 by 27 inches.)

Next he saw a picture of the Earth from space; he recognised the shape of Africa, and Anouxia said the word "*Earth*". An oddity of the pictures of planets which he saw in the screen, is that they show no cloud-pattern, even where one is present at all times; one could not photograph Earth from space without recording the characteristic delicate fleecy veil of white clouds; and Janos, seen later, likewise had no clouds: it may be that the photo-technology employed was such that it did not show clouds, so that the planetary surface was sharp and clear; this would be an advantage in their accustomed task of planet survey. John is convinced that it really was the Earth, and not a model.

The picture of the Earth was followed by one of the Moon with its characteristic markings and craters, which he knew from pictures in books. Next, a brown planet with markings on it, which he did not know, but from his description it could have been Mars. Then a whole series of planets of many kinds, mostly golden-brown in colour where the sunlight caught them, went streaming slowly by; these were all unfamiliar. Once there was a large brown planet with crater-like markings; it looked like a half-moon, the sunward edge brightening to a golden hue.

Next, he saw the curved edge of a planet quite near; and looking beyond it, he could see a small planet or satellite much further away. John says that the '3-D' effect of the screen was very pronounced; he was very conscious of the nearness of the one planet, compared with the more distant one. All the films shown to our witnesses had this enhanced stereo quality; and they remarked more than once on the extremely real and life-like quality of the pictures, so that, as he put it, you felt that you were actually present, and not just watching a film. He was unaware of anything outside the rectangular boundaries of the screen area, and unaware, in the later sequences, that it had boundaries. Leaning forward and looking into the large horizontal screen, he was right there in space, experiencing for himself.

Then there was a very large pinky-red planet, which filled the screen so that the top and bottom edges were cut off; but he had a sense that it was very far away. He said it wasn't because it was near that it seemed so big. (This may have been something he was told; John's memory of speech over this section is still patchy.) He said that the pinky colour was variegated by hundreds and hundreds of vague smudgy squiggles of a colour still pink, but a deeper pink. As he watched the great pink globe, the camera viewpoint swung slowly round it, as if the craft carrying the camera were orbiting, for about half an orbit. Because of its great size, and his sense that it was far away, I have wondered whether this was not a planet at all, but a cool star – a 'red giant'. The 'squiggles' could well be a variant on the granular structure of our own Sun's photosphere.

All the planet-images that he saw behaved in the same way: the planet would appear abruptly, somewhere on the screen, not always centred, and then very slowly drift to the right, becoming smaller and smaller, giving the feeling that it receded, until it moved off the right-hand margin of the screen. John said he had a feeling it wasn't the planets that were moving; it was he who was moving past them (notice again the sense of subjective, rather than vicarious experience).

In this connection, he felt at one point that the whole part of the film that showed planets was being shown in reverse; perhaps he was told this. If this was so, Earth came at the end of the story, not at the beginning. The identity of the

Pictures in the Navigating Screen

ridged planet he saw before the Earth was never revealed to us.

One should remember, throughout this pictures-in-the-screen sequence, that he was seeing the pictures upside-down compared with Anouxia's view of them, since they were on opposite sides of the table. With these 'space' pictures of planets, it probably did not matter which way up you viewed them; but later, when normal views replaced the space pictures, John was seeing them right way up, so the film was evidently presented correctly to his view.

Next there was a golden-coloured planet, with vague darker brownish shapes upon it, arranged as it might be continents on the Earth; but the boundaries were fuzzy and indistinct. It seemed to him about the size of a golf ball. Grouped around the planet were five smaller bodies, round in shape; John said of them: "They're much brighter – more of a white or silver colour. And they're more the size of a sixpenny piece to a shilling piece." This would be from five-eighths of an inch to nearly an inch – 16 to 24 millimetres.

The whole group, as usual, drifted across to his right, receding; but apart from this they kept their orientation and arrangement steady, except that at one point, quite suddenly, the whole group rotated through an angle, anti-clockwise; then stopped again: presumably somebody had adjusted the camera.

The screen had been gradually darkening; by now it was really a very deep blue; in the background he could see many stars – pinpoints of brilliant silvery light, which did not twinkle; of course in outer space stars do not twinkle – this is an effect of the Earth's atmosphere. John did remark that the stars seemed more stationary, compared with the planets which drifted across the screen; he said, carefully: "If they're moving, they're a lot slower".

Later, a cluster of many objects drifted past; these were not round, but irregular, craggy shapes, somewhat elongated: the long axes of all of them were parallel to each other, and also parallel to the apparent line of drift, suggesting a real movement, as distinct from a camera displacement. Some of the bodies were clearly nearer than others. They were brown in colour, with golden edges to them, where light caught them from one direction; there must have been a star near enough

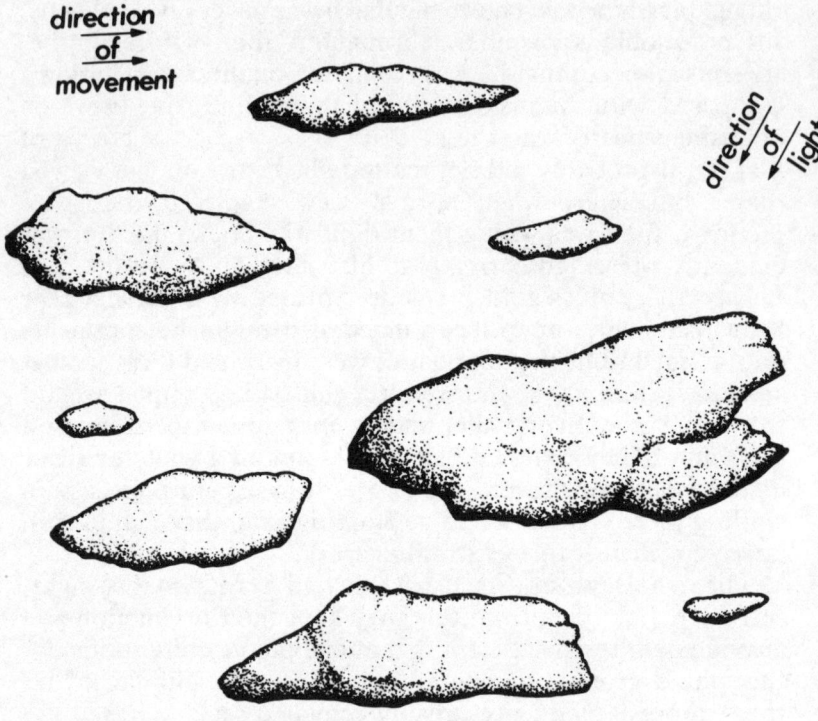

15 A flight of asteroids or planetary fragments, seen by John in a film

to illuminate them. Anouxia said: "*These are too small to live on*"; indeed, so much can be deduced from their non-spherical shape: a planet large enough to retain an atmosphere must be nearly spherical.

Many of our asteroids in the Solar System have this irregular, craggy appearance; they are not massive enough to crush themselves into a ball by their own gravitation, and certainly not big enough to hold an atmosphere which could support life. Most probably, these craggy shapes that John saw were, like our asteroids, fragments of a broken planet. This would account for their travelling in company.

One must remember that the Janos people were all the time

Pictures in the Navigating Screen 113

looking at planets, not just to satisfy their scientific curiosity, but as possible homes. These pictures were probably taken at different times during their voyages of exploration through space, looking for a place to live when the time came for them to leave Janos.

* * *

Finally, they came to a picture of Janos and its two moons. There was a group of three bodies, forming a loose triangle on the screen: they appeared to be much the same size; but in fact, from what we know from other parts of the story, this must have been a picture taken when, by chance, the two little moons and the planet were almost lined up – otherwise they could not have been seen together in one picture with almost equal apparent sizes.

Clearly the camera was beyond the outer moon (which we think is called Sarnia – pronounced Zarnia but spelt with an initial S, like Saton, the inner moon), and looking inwards past Sarnia and Saton to the planet Janos in the background. No other interpretation of the picture is gravitationally feasible. It follows that the outer moon is smaller than the inner moon Saton, since they appeared the same size, but the outer moon was nearer the camera.

(On one occasion, Uxiaulia remarked to Frances, speaking of Saton, *"we could see it from the ground"*. The name Sarnia was referred to several times by Uxiaulia in the same context; he gestured upward while saying it. Frances was not quite sure what he was referring to, except that it was a heavenly body of some kind; it could have been the name of their Sun, but from the way he spoke of it, it seems more likely to be the name of the outer moon, for which we have no other name.)

Anouxia put his silver-gloved finger on the image of Janos, saying: *"This was my people's home"*. Then he pointed to the inner moon, saying: *"This one was too close"*. Saton was, indeed, too close to the planet for stability; the story has been told in the Prologue of this book. That Saton was too close to Janos was, in fact, the main cause of the disaster which destroyed the Janos people's planetary home.

The planet Janos, as John saw it in this, and later in closer views, was a greeny-brown colour, varied with patches of blue; these blue areas were water. There are many lakes, and some large areas of sea or ocean. This was the only planet other than Earth, among the many that John saw, that appeared to have water on it. The two moons were the usual golden-brown. He saw no white areas on Janos, indicating snow or ice, at any time; Uxiaulia told Frances: *"On Janos it was always warm"*.

The film now cut to a much closer view of Janos, so that it more than filled the screen; but part of the curved edge could be seen. It appeared to be turning very slowly from left to right; but the probability is that the effect was produced by the orbiting of the spacecraft which carried the camera. Indeed the view of the planetary surface gradually came nearer as it apparently turned, indicating that the craft was spiralling in from space towards the surface.

The seas and lakes could now be clearly seen. There were low hills, but nothing spectacular in the way of mountains. In this view of the planet, filmed before the catastrophe, John had an impression of greeny-brown areas which looked fertile; but it was too far to make out detail. Later, he said that some of the green areas – it was a dark bottle-green – did have the look of forests seen from a great height.

There was another abrupt cut in the film, to a much lower altitude: it is a pity that he has not fully recovered memory of what was said to him by way of explanation; he has a feeling that something was being explained at this point, but cannot recall it. When he first described this scene under regressive hypnosis, we did not yet know about the rockfall; we later deduced that the film sequences from this cut onwards were all taken after the catastrophe.

As the camera-bearing craft spiralled in, the ground detail became plainer. The apparent or false continuity, which seemed to bridge the cut, making a seeming transition from one spiralling-in scene, before rockfall, to another spiralling-in scene, after rockfall, confused and misled us for a time; if we had had the full verbal explanation, we would have understood sooner what was happening.

The new, much closer scene made John remark that the land areas seemed covered with thousands of tiny bumps, of a greeny-brown colour. He was still at a great altitude; as the camera came gradually lower, he began to realise that the land surfaces were covered with a dense layer of great rocks, which had appeared tiny because of distance.

Now he was near enough to have seen forests, towns and cities; but there was nothing but the endless dry sea of loose rocks. Not then knowing of Saton's breakup, I tried to understand how such a weird planetary surface could have developed; I could not understand how millions of loose rocks, piled in a completely disorderly way, could come to cover a planetary surface. Later, of course, it was at once clear; had I been thinking more quickly, I would have made the mental jump in one go, from *"this one was too close"* to the rock-strewn landscape. It was not until Frances, in my presence, on a later occasion under hypnotic regression, re-lived her viewing of the film which, from the ground, showed great rocks falling from the sky, that the 'penny dropped'. 'This one', meaning Saton, was indeed too close.

The blue areas of sea and lake were still there: rocks must have fallen equally over water and land; but in water, beyond making a lot of tremendous splashes, they would have sunk without permanently altering the appearance of the seas and lakes, except perhaps at the coastline. John, under hypnosis, was clearly puzzled by the juxtaposition of what he thought of as a 'desert' landscape with large areas of water – though this does occur on parts of the Atlantic coast of Africa, where, in some regions, desert rock and sand go right down to the ocean beach.

In one place, when the camera view was becoming really low, John said he could see, on the margin of a lake, some reeds or similar vegetation; but they seemed dead and brown. Nowhere did he see any trace of civilisation: no towns, large buildings, roads or railways, or any sign of habitation; and, as he passed over the dark side of the planet, no glimmer even of artificial light. We all felt a keen sense of disappointment; we did not know, at that stage of the investigation, that he was looking at a world that had been very much alive, but had been crushed out of all recognition by the battering of the rockfall.

Finally, the camera-view of John's present story came so

near the ground that he felt the sensation familiar to all of us when coming in to land at an airport: just before touchdown, the ground seems to be hurtling past at tremendous speed, though the craft is now travelling more slowly than at any time in the flight.

John's camera viewpoint came to a stop, perhaps three hundred feet above the ground. The view was directly downwards, into an untidy jumble of greeny-brown rocks – the occasional big one, but mostly small, including a lot of shattered fragments; he could see a fair amount of heavy, gritty dust, no doubt the result of big rocks falling upon earlier-fallen rocks, and smashing and pulverising them – though there is another, more sinister explanation for the dust.

The greenish tinge on the rocks is a bit puzzling: it could, of course, be inorganic; but this part of the picture from other, internal evidence, was probably filmed as late as several months to a year after the end of rockfall, not long before the fleet left for Earth; and it is possible that simple forms of aquatic life in the deeper waters could have survived the radiation holocaust, and were beginning to film over the rocks with unicellular forms of life: rain would wash the rocky surfaces clean of radioactivity after a time. Eventually, no doubt, life, perhaps a much-mutated life, will return to the land on Janos; but it will be a very long time before people will be able to live there safely – perhaps hundreds of thousands of years.

As he watched the film, John was startled to see movement in the centre of the patch of rocky wilderness that the camera covered, looking straight down. There was a slow stirring among the rocks almost as if something big were pushing up from below. The rocks and debris heaped up into a restless mound; and then the bigger pieces began to slide, roll and tumble slowly outwards, away from the centre of disturbance.

Listening to John's puzzled description, in deep trance in the hypnotist's consulting room, was an eerie experience: what new surprise was this devastated world going to spring on us? People under hypnotic regression are apt to talk very quietly, making problems for the microphone and tape recorder; in this passage, his voice becomes unusually faint

Pictures in the Navigating Screen

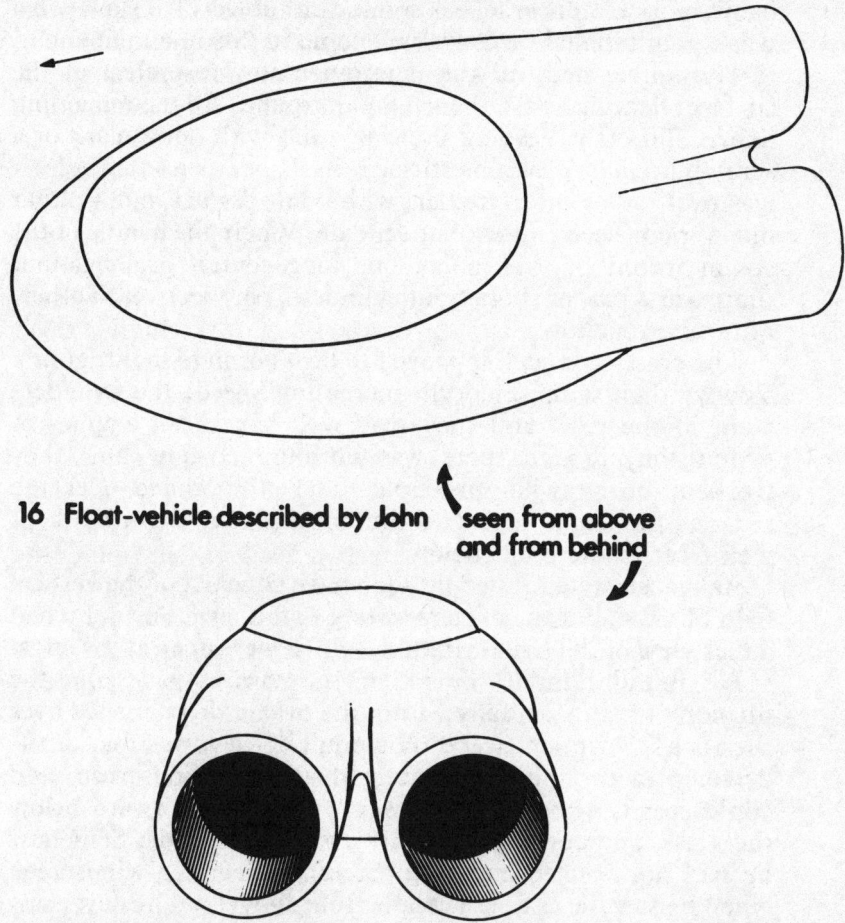

16 Float-vehicle described by John seen from above and from behind

and indistinct, as if he were far away. I am putting together this account from what he had told me on various subsequent occasions, as his amnesia slowly dissolved and memory came back, first the vision and later the sound – and speech later still.

The rocks seemed to burst asunder in slow motion; and a round, smooth object appeared, made of some shiny dark metal. As it rose steadily out of the ground, the rocks and debris slid away from its smooth domed back, then cascaded off its outer curve, rolling and falling, all in slow motion. The 'slow-motion' appearance is significant; objects fall slowly in a weak gravitational field: we will come to this in a moment.

The whole body of the newcomer now rose clear of the surface, floating above the irregular ground. All the remaining debris slid off it, leaving it clean – and with not a mark or a scratch upon its gleaming surface. In shape, seen as a whole, it was oval rather than circular, with a low domed roof – rather like a spoon seen from the under side. Where the handle of the spoon would be, instead of one long central prolongation, there was a pair of short stout cylinders, parallel to each other, not quite touching.

The craft – for such it proved to be – began to move, at first slowly, then with smoothly increasing speed; the cylinders were in the rear, and they may well have been engines of propulsion, though there was nothing to show how they worked, or by what principle. The craft glided over the surface, picking up speed; it did not rise at all high, but kept well clear of the rocks below.

As it accelerated, it would soon have gone out of the vertical field of view; but the camera swung to follow it, and John had a rear view of the twin cylinders, which were open at the ends.

He found himself, or rather the camera gave him the illusion of being, actually within the oval craft, as it sped over the rock-strewn landscape. His point of view was that of the driver of a car; he was looking through a windscreen, and could see the 'bonnet', as it were, projecting forward below the glass, and curving smoothly away downwards. Somehow he had not noticed anything corresponding to a windscreen when he saw the craft as a whole from above; but he never saw it from in front.

The craft was now moving fast; and suddenly he noticed that it was approaching a very large and curiously-shaped wide tunnel-mouth: the actual entrance to the tunnel was not just a hole, but a well-constructed piece of engineering. It probably showed some signs of damage; but John did not mention any.

The shape of the opening, and of the tunnel section within, was a flattened diamond with rounded corners: the middle of the roof was gently rounded, then on each side it descended in a sloping line towards the widest part, midway between roof

Pictures in the Navigating Screen 119

and floor. The line swept round in a fairly tight curve, continuing as a line sloping inwards towards the middle of the floor. In its lowest part, the surface of the floor of the tunnel was flattened to form a roadway, instead of being curved to mirror the curve of the highest part of the roof. A difficult shape to describe; but I have provided a diagram.

The oval craft entered the tunnel at some speed; and John noticed that the tunnel section was a great deal larger than was needed to allow passage to this particular vehicle. Clearly it was designed to take a much larger craft, of a peculiar shape: it did not occur to him to think of a flying saucer, airborne, like the ship he was watching the film in.

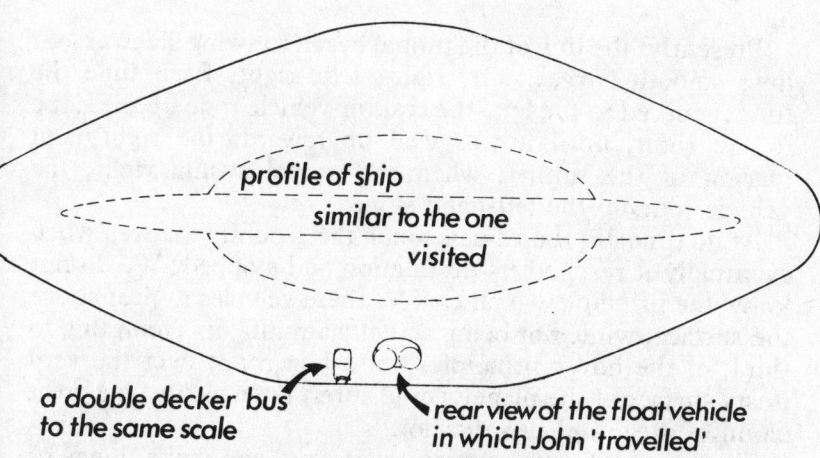

17 Section through one of the great tunnels which lead down to the underground shipyards

CHAPTER TEN

Underground Encounter

THE TUNNEL SLOPED down into the ground fairly steeply: for a time it was straight, and the oval craft, in which John was vicariously travelling, accelerated until the walls simply flashed past. There was enough light to see where you were going; John's feeling was that the vehicle itself carried headlights.

Presently, the line of the tunnel began to swing sideways, in long smooth curves, left, right, left, right. Each time the tunnel curved to the left, the craft or vehicle rode up the slope to the right, until it was well up towards the right-hand margin of the tunnel; when the tunnel swung right, the vehicle rode up the left-hand slope.

At no time did the vehicle touch the ground, not even when eventually it reached its destination and stopped. We do not know the principle which enables these vehicles to float above the surface, while not being actually aircraft; it is tempting to think of the hover principle, but a hovercraft over the very dusty surface of the planet would surely have blown up a great cloud of dust, and this did not.

Since the ability to control gravitation, and make things or people float up or down, is a known part of the Janos technical repertoire, one need not look further for an explanation, even though we do not yet know how it works.

Soon the vehicle was travelling really fast; the tunnel continued to slope down, and by this stage it must have been a long way below the surface. John became conscious of a whirring noise.

Presently it slowed, and eventually came to a halt, still floating. John found himself (under hypnosis he was half convinced that he was actually there) at the entrance to a vast, gloomy cavern, the roof of which, almost lost in the darkness above, was supported by many massive columns, cylindrical in shape, hewn out of the solid rock. The top of each column expanded, trumpet-like, to meet the roof.

At first he reported that he was in complete darkness; then as his eyes became accustomed to the gloom, he began to make out something of his surroundings. He said the place was dirty; there was a lot of gritty dust on the floor. He became aware of a group of half-a-dozen people who shuffled slowly and wearily towards the front of the vehicle, coming from the left. They were carrying something between them – something long and heavy; as the group moved into the lighted area in front of the vehicle, he saw that it was a crudely-fashioned wooden coffin. It was not shaped like ours; just a box.

The people – he could not tell if they were men or women, and it scarcely seemed to matter – seemed not far from death themselves. They shuffled along slowly, in a dejected way, "as if they had given up" he said. Accustomed to the gleaming silver uniforms of the spaceship crew, he found their clothing strange – yet strangely familiar. They were like monks; a monk's long-skirted habit, black, with a deep cowl or hood over the head, half-concealing the face.

It was a merciful concealment. As they came fully into the vehicle's headlights, John could see their faces. He is a sensitive person; and under deep hypnotic trance it really upset him badly, the first time he saw them. At first he said they were "old"; but this could not account for his horror: old people are not frightening in that kind of way. Then he corrected himself, saying they "look old"; they were prematurely aged. (Remember that at this stage, we did not know about the radiation sickness; this was the first hint we had of it.)

When Frances saw the same kind of people in another film, in other circumstances, she also reacted in the same way – a mixture of pity and revulsion, with compassion very strong in Frances. She called them 'Oxfam people'; she said they were 'like lepers'.

The faces looked dead; the eyes were dim, or in one or two, had gone altogether, leaving them blind and groping. Their teeth were all gone, leaving the cheeks sunken in. The hair, originally fair, was limp, lifeless and straggly, like damp straw, where it showed beyond the cowl. The fingers were deformed, swollen and claw-like, with big lumps on the knuckles; though clearly they could still carry a load. No one went out from the vehicle to help them.

The bearers and their load passed under the front of the floating vehicle, and John lost sight of them. Presently, they reappeared without their burden, returning quietly to the shadows whence they came; but separately, not as a group. From what we learned later, we realised that they had loaded their dead into the lower freight hold of the vehicle; and that this was a routine visit, to collect the corpses and take them away for disposal. They were like living corpses themselves; and unspeakably sad and forlorn.

John had always had a tendency to forget he was merely watching a film; under hypnosis he kept talking as if he were actually present at a real event; the vivid realism of the pictures helped the illusion. At this point, he had a moment of illogical panic, thinking he was going to be left there, deep underground with these horrible people, and never get out again.

Presently, to his relief, the vehicle began to move again, at first slowly, then gaining speed. It returned by way of the same tunnel to the surface and daylight; emerging from the tunnel mouth, it sped across the rock-strewn surface. Another film cut transferred his viewpoint back into the spacecraft overhead, so that he was once more looking down from a low altitude. The craft below went out of the picture; and the film ended.

* * *

I have told, in the Prologue, how many people, caught by the unexpectedly early beginning of rockfall, had made for the tunnels, if they were near enough, and had made their way – it must have been a long way – to the shipyards deep underground, where they expected to find safety and supplies.

But their death certificates were signed, from the moment, seen by Frances, when the first nuclear power station exploded, triggering off all the others, right round the planet, in a giant chain reaction. They had a matter of months to live, at most. Nothing could be done for them. The people in the ships safety in orbit did not know what had happened to them, or even if there were any survivors – certainly there would be none on the surface.

When they knew the truth, when the rescue ships came, lifting heavy rocks to clear the choked tunnel-mouths, the ship people suffered a deep psychological trauma, which leaves its scars to this day. They could not help the dying people, beyond organising as best they could for their relief, without themselves running the risk of picking up radioactive contamination which might spread to the ships in orbit.

Someone, in a thoughtful moment, devised the monk's clothing, to give them better protection against the lethal dust; perhaps they did it themselves. Our witnesses were told that this was not their normal clothing, but a special garment designed to keep the dust off them; it was much later that Frances recalled the film of the happy, carefree times before the disaster, which has given us our only glimpse of normal private life on the old Janos.

We have wondered sometimes why, knowing that the doomed people underground faced a certain but lingering death, they did not give them a merciful euthanasia, rather than let each individual life drag out its slow and inevitable end. One can only imagine that the idea of mercy killing is just not in the Janos people's philosophy.

* * *

We must look, before we leave this terribly sad episode, at a technical problem: how did the oval craft manage to force its way up from below, through a deep layer of rocks, some of them probably weighing thousands of tons, without showing so much as a mark or a scratch upon its darkly-gleaming polished metal body?

One must assume that it was engaged on the task of clearing a choked tunnel-mouth, coming at the job from below, having entered the tunnel system by another route. The entrance that John went in by was already cleared; perhaps it had been more lightly obstructed, for John had the impression that it ran into something of a hillside, where the rocks might not have accumulated – though he saw, in the later part of the film, no part of the land surface which was not rock-covered. Even a small moon, broken into fragments, will yield an astonishing quantity of rocks.

The clue, as to how these enormously heavy rocks were lifted by a vehicle of no stronger construction than an ordinary bus, lay in John's impression that the whole business of the emergence of the vehicle from below took place in slow motion. As I have already said, a heavy body will fall slowly in a weak gravitational field: we know, from many demonstrations, that the Janos people are able to control gravitation and therefore weight; their own ships and vehicles 'float', and John and his family were themselves floated up into the spaceship and back down to the ground when they left – and we learned that the ship's own elevators work on the same principle; they have no cage – you simply step into the lift shaft through a doorway and float slowly up or down.

I think that what the oval vehicle did was to create around itself a zone of feeble gravitation. The vehicle itself was given a slight lift – a very weak negative gravitational field. Theoretical physicists will, I am sure, inform me that negative gravitation is an impossibility; all I can say is, that through John's eyes and excellent visual memory, I saw it happen.

We know that these zones of controlled, even reversed gravitation are closely defined; when the family were on the road by their car, ready to go up into the spaceship, Frances remarked that she did not begin to feel the lifting sensation in her body until the beam of light, which at first fell as a bright circle on the ground in front of them, moved back until the group of people were within its illumination. Either the projector was adjusted, or more probably the ship moved as a whole, just a few feet.

Whether the light beam had any functional connection with the 'anti-grav' principle I do not know; it may have been merely a marker. Other cases have been reported of people floating up a beam of light. At least one of the ship's elevators was at times associated with a vertical beam of light; though when John used one to visit the engine room and the bridge, its interior was almost totally dark. If we are right in concluding, as we did later, that the same elevator was used on both occasions, then it is sometimes light and sometimes dark, for a reason we have missed.

Returning to the vehicle which came up through the rocks: there must also have been some force which prevented dust and debris from adhering to its upper surface; John described it, as soon as it emerged, as clean and smooth, with a gleaming, polished surface of a dark colour. An electrical repulsion field may have been employed.

CHAPTER ELEVEN

Departure from the Spaceship

AT THE END of Chapter 6, we left Frances with Gloria and the children, in an upper room of the spaceship, waiting for John.

John did eventually appear; he was conducted by Anouxia from the navigation room, by way of another moving ramp which led up through a doorway, wider than usual, which gave entrance to the upper room. Anouxia left him there, and went back down the ramp, no doubt back to the navigation room; the imminent departure of their guests meant that the spaceship would soon be on its way.

The way the ramp entered the upper room was strange; at the doorway, the ramp surface was still some three feet below the general level of the floor, and it continued to slope up into the room, until it reached the floor level. When Frances and the others first saw John come through the doorway, they saw only the upper part of him, coming up from a lower level; and when John first saw them, they were above him.

I must interpolate here a circumstance about doors and doorways which we found unexpected; at first it was difficult to interpret and caused some confusion. As far as we were able to discover, it applied to all the doors in the spaceship.

The doorways were not rectangular at the top like ours; the corners were rounded and the lintel slightly arched. There was, we think in all cases, an actual metallic door panel, which opened by sliding sideways into a recess in the thickness of the wall; although one sometimes passed a closed door, generally they were left open.

So far, nothing unusual. But if you stood in the corridor, looking through the open doorway, you did not see the interior of the room: you looked at nothing. This is not at all the same thing as looking at a black door; the effect was an

Departure from the Spaceship

impenetrable inky blackness, like the interior of a subterranean cavern when the guide turns the lights off to frighten you.

Similarly, if you stood inside the room, looking out through the open doorway, you did not see the corridor: you saw nothing.

Our family at first hesitated to penetrate this veil of total darkness; but seeing the spaceship people do so without hesitation, and vanish, they did the same, and instantaneously found themselves in the room, without any sensation of having passed through anything. When, as happened more than once, they passed from a brightly lighted corridor into a darkened room, the instant they were past the plane of the open doorway, they were in the dark; there was no spill of bright light following them into the room through the open doorway, and lighting up the room. When Frances came to leave her medical room, the door was closed; as she approached it, she saw the metal door-panel slide away quickly to the left, leaving an opening of inky blackness, through which she had to walk.

When Frances, Gloria and the children saw John come up the ramp into the upper room where they waited, he burst abruptly into their vision out of a dense black door-opening; by now, they were getting used to this bizarre effect. Somehow, the Janos people have invented a means of preventing light from going through a doorway, without hindering the passage of people or other material objects.

A convenience, no more; but it illustrates well their mastery of science, that they can play such tricks with light. But, of course, their really spectacular tricks, reported elsewhere in UFO literature, are the ones that must depend on a total control of gravitation and inertial mass – enabling their spaceships, for example, to change instantaneously from very high velocity to dead stop; or, while travelling very much faster than any terrestrial aircraft, to change direction abruptly, turning on a point. Either of these manoeuvres would smash up a terrestrial aircraft, killing its crew. We have seen numerous applications of this gravity-mass control in the spaceship described in this book.

The upper room (we called it that, because it must have been on one of the highest decks of the spaceship), in which the family met together before their departure, has been described at the end of Chapter 6; the most prominent, and to Earth people the most puzzling feature of the room was the bright vertical cylinder which we have already mentioned.

We had a lot of difficulty with this bright cylinder. There was some difference of interpretation between Frances and John, concerning the cylinder: Frances described it as being like a piece of a huge fluorescent tube, five to six feet in diameter, extending from floor to ceiling. Most of the time it glowed brilliantly, filling the room with light; but at certain times it was dim. They agreed that the intensity of the light was uniform from floor to ceiling; there was no falling off of the brightness from top to bottom. They were agreed about its size.

The difference between them lay in the fact that Frances thought of the cylinder as a material object – a hollow cylinder, in fact, made of translucent glass or plastic. John, on the other hand, persistently referred to it as a 'beam'; he said that he thought of it as a beam of light projected from above the ceiling, and passing through a circular hole in the floor. He did not think of it as having solid walls. Perhaps he associated it, in his mind, with the beam of light which had mysteriously lifted them up into the spaceship, when they first arrived about fifty minutes earlier; the reason for this association will shortly become apparent.

When John came into the upper room, the bright cylinder was in front of him, not quite directly in front, but a couple of feet or so to the right. As he was carried up the remaining part of the moving ramp, which extended into the room where the others were waiting, he came up to their level; they were all standing with their backs to the bright cylinder, partly to face the doorway where they expected John to appear, partly, no doubt, to shield their eyes from the dazzling light of the cylinder, which was so bright that if they looked at it directly, they could not see the room at all clearly, as their eyes were confused by the brightness.

The four members of his family awaiting John stood in a row, as he first saw them: from left to right, Tanya, Gloria, Natasha, Frances. When he first came up, Anouxia was still with him, standing on his left; the ramp was wide enough for two people to stand side by side, and they had come up this way: the doorway, which they passed through when they entered the room, was unusually wide, to allow the ramp to pass through it; it was of the usual rounded-corner, slightly arched type. Like all other open doors, you could see nothing through it, until you passed through it yourself.

Two or three of the silver-clad ship people stood around, one of them directly behind Gloria, as John arrived. No one spoke, except that Natasha said: "Here's Daddy".

One of the ship people presently gathered them together and led them round to the back of the bright cylinder, on the opposite side to the one they had had their backs to. Here, Frances says, there was an opening or doorway leading into the interior of the cylinder; they all went in, together with their silver-clad guide, and stood in a group within the cylinder. The light dimmed right down: either way they could not see very clearly; but they were all conscious that they were standing upright, all standing at the same level as if there was a floor under their feet – but there was no floor; they were suspended in space.

They floated slowly down the cylinder, still unsupported, still standing upright in a normal way as if they stood on a floor. It was just like going down in an elevator – but much smoother than any Earth elevator – except that there was nothing under their feet. Both Frances and John are quite sure of this. They felt quite normal; not dizzy or unbalanced in any way. There must have been a fair amount of space; because they agree that they were not crowded together, and there were four adults and two children altogether.

Quite soon they came to a gentle stop – no jarring – and found their feet on a solid deck. Guided by the ship man, they passed out through a doorway of the usual kind, and found themselves on the balcony – the same balcony that John had stood on, the same balcony which they had all come up to by the moving ramp, when they first arrived in the huge circular engine room.

They all went down the moving ramp – this time it carried them downwards – and walked across towards the centre of the circular deck, a considerable distance. As they did so, they could see that through every doorway in sight, people in silver suits came streaming out, scores of them, to see them go. Half a dozen of the people mainly concerned with their visit, including some women, waited for them by the rectangular inner hatchway of the airlock, which was open. The rest of the ship's company kept well back.

The two groups met together, and formed a single ring of people, silver-clad ship people and Earth people in their ordinary summer travelling clothes, mingled together. There was a sense of occasion; almost of ritual.

Anouxia again spoke for them all. Addressing the visitors, he said: *"It is time for you to return to your car now; we are in position"*. No one answered for the Earth people; but all were filled with the same reluctance to leave. John said afterwards that he felt it would be better if they were able to stay longer with the Janos people.

He added that Anouxia understood their unspoken thoughts, and smiled; but he said firmly: *"You must return; we will see you again. When you see us again, you will know us"*. Frances confirms this. Anouxia went on to explain that they must not stay too long, or people would become anxious about them, and enquiries would be made.

Someone produced a tray with glasses of a colourless, fizzy drink. *"Would you like a drink before you go?"* They all accepted, and John said: "Is it alcohol?" and was answered: *"No; but it will help you to forget"*.

John asked: "Why do you want us to forget?"

"Because if you remember everything immediately, you will go around telling everyone, and it will cause you much trouble; many people will not believe you, and others will try to exploit you. You will remember everything in time; but it will be some time before it all comes back." And this is how it was; except that we helped the return of memory by hypnosis: even now, not everything has been remembered.

Frances says that she did not like the taste of the drink much – it tasted milky; but all the grown-ups drank theirs down. Natasha did not want hers; and Tanya drank a few sips and put hers down. Akilias, the woman who had looked after Natasha, said: *"It will not matter if she remembers; because she is so little, no one will believe her"*.

It was time to go. Anouxia, again, they felt, acting for all the ship's company, shook hands with John; then he kissed both the little girls. Next he took both Frances's hands in his, and

Departure from the Spaceship

kissed her on the cheek; then he did the same with Gloria. "*I promise you that you will see us again*", he said. "*You will be all right; we shall be seeing you home.*"

John, practical as ever, said: "Where is the car?" – for he had earlier seen it standing near the airlock hatch, on the main deck. Someone led him to the edge of the big square open hatchway, and he looked down: directly below him was the car, its white paint catching the light from the spaceship; it was neatly parked close by a tall hedge which John could just make out. John said he could see the whole length of the car, and from this he judged that it must be at least thirty feet below them; I think, from other considerations, it was probably nearer fifty feet.

They all looked down; it seemed very dark on the ground (doubtless because the ship itself shadowed the ground), and a long way down. The children looked; but quickly lost interest in the car; they were soon gazing around the huge circular room, with so many people.

When it was indicated to them that they should step out into unsupported space from the edge of the hatchway, naturally enough they hesitated, despite their recent rather confused recollection of the bright cylinder. A beam of light, but not a very bright one, did in fact shine down on to the ground, making a circle of light quite near the car.

One of the men volunteered to go down with them, "to show us that it was O.K." (John). This man – it was neither of the two they knew well, but another one – stepped fearlessly forward from the edge of the hatchway, as if he were walking out on to a solid floor; but there was nothing under his feet. He walked, or glided, forward until he was well out in the middle of the hatchway, still unsupported, but standing normally, as they had done in the bright cylinder.

The visitors therefore plucked up courage, and walked out after him, until the six of them were standing together, still level with the main deck, but standing on air.

Then, presumably, someone pressed a switch somewhere; because the whole group of six began to float quite slowly down towards the ground, keeping their level formation. As they emerged clear of the airlock, through the lower hatchway, they felt the cool night breeze blowing gently in their faces. The downward movement was slow, but steady; presently they touched ground, so softly that their knees did not even bend.

They were a few yards away from the car, on the driver's side. They were themselves standing on a pathway, beside the road; the place was, in fact, a small car park, rarely used at that time of night, just outside the town of Faringdon.

The light beam still shone on them; but now it moved back from them, over the roadway, just a few yards. They thought the spaceship as a whole moved back a little, causing the beam to move off them.

The silver-clad Janos man, who was still with them, walked slowly backwards, away from them and towards the circle of light cast by the beam on the tarmac road. As he walked, he said: "*You will remember none of this: you have been driving*". Then he said it again. As he passed backwards into the circle of light, he began to float up slowly, all the way up into the spaceship; and the hatch closed.

As they stood watching, the spaceship lifted away, then came lower again; and they could see, through rows of lighted windows, the heads of many of the ship's company looking down at them in the little car park – "as if they were waving goodbye".

The ship finally lifted quickly away, and was soon lost to sight. They turned towards the car, and found everything in order; the lights and ignition were off, and the keys were in their place in the ignition lock. They all got in from the driver's side, since the other side was too close to the hedge to open the doors; John drove away, and at once they were in Faringdon. From this point, the 'real story' rejoins the 'cover story' as told in Chapter 1; readers who wish to round off the story may like to re-read the end part of that chapter, from page 15.

* * *

The hedge in the car park is important to the mechanism of the 'cover story', and I will deal with it here.

You will remember that the unreal and interminable drive through a narrow lane, with high hedges close in on either side, had an element of repetition: John, who, as driver, was perhaps most directly involved in the 'cover story', remarked to me much later that a pattern seemed to repeat itself endlessly; he also said that the two hedges that they drove between, one on either side, were as it were mirror-images of each other.

I think there is no doubt that the five were, throughout their stay in the spaceship, under hypnotic control, which began when they first saw the rotating circle of coloured lights. John, by the time we finished his prolonged series of hypnotic sessions – he had ten, some of great length – was fairly experienced in hypnosis, from the receiving end as it were; and he gave it as his considered opinion that he was hypnotised throughout the spaceship visit.

Undoubtedly the fizzy drink also contained a hypnosis-predisposing drug; such drugs, Geoff told me, are well known on Earth: the post-hypnotic suggestion, that they would remember only that they had been driving, would be much strengthened by the drug, which they were told would help them to forget; and no doubt it prepared their minds for the artificial pseudo-experience of the narrow lane. The children had little or none of the drink: their memories of the whole incident were clear – though there was probably a degree of mild amnesia in Natasha; her memories have a habit of re-appearing in stages, like those of the adults. As far as we can tell, the children have no recollection of the 'cover story' drive. It is also noteworthy that Gloria, whose amnesia of the spaceship visit remains almost total, had a very clear memory of the one part of the story that wasn't real.

It is not clear whether the visual background to the 'cover story' was deliberately arranged, in detail, by the spaceship people, or whether it was more or less an accident; I personally incline to the latter view. I think the adults were hypnotically pre-programmed to accept the suggestion that they would remember driving for about fifty-five minutes, to account for the loss of time; this extra driving time was to be inserted into their real drive to give an illusion of continuity; remember that Anouxia, in his speech of welcome on the balcony, said: *"We will replace you back in your car, exactly as if you had never stopped"*.

In fact the joining together, the sewing-in as it were, of the artificial memory-insert was very smoothly done. The illusion failed to be convincing for two reasons: one, which the spaceship people possibly were not aware of, was that the family travelled this road regularly, and knew every inch of it, so that even a slight departure from the normal would have been noticed; the other was the singularly unconvincing visual pattern of the narrow lane with the close-set tall hedges. They knew very well that there was no such lane anywhere on their route.

My own interpretation, based on what Frances and John have told me, is that the 'lane' appeared narrow because their car had been parked by the ship people, when they set it down in preparation for the family's departure, close by a short length of tall hedge. By some trick of suggestion, they 'remembered' seeing the hedge on both sides of them, equally close; so that they would seem to be driving along an extremely narrow lane, so narrow that they could not have passed another car. Fortunately this problem could not arise, because the experience was unreal.

John, Frances and I revisited the car park, and were able to pin-point accurately the position of the car as it was set down by the hedge; we were also able to work out, from visual angles, just where the beam of light had set them down by the roadside, a few yards away. There are a few buildings around; but they are unoccupied: so no one would have seen the incident from a window. There was, of course, the possibility that, even late at night, a car might come along the road: no doubt the spaceship crew had checked carefully along the road in both directions, before setting them down; but since they were on the outer edge of a town, there remained the possibility that someone might come along and have a big surprise.

Both John and Frances identified the hedge as the one in their 'cover story' drive; the illusion of driving for nearly an hour along the narrow lane, closely hedged on both sides, was built up out of the very brief glimpse – it cannot have been more than a few seconds – of the short piece of real hedge, on one side of the car only, as they drove out of the car park, on to the road. They were inclined to think that some visual detail of trees on the far side of the main road had also been worked into the illusion.

I am inclined to think that the choice of visual material by the Janos people, out of which to create the illusion of the prolonged drive along a narrow closely-hedged lane, was largely opportunist; the material presented itself at the moment at which John began to move the car forward out of the little wayside car park on to the road.

The family were hypnotically pre-programmed to accept this visual material as the basis of a long drive, to account for the time lost; but the actual choice of material must have depended on what was available.

As an illusion, it was unsuccessful and unconvincing, and failed of its main purpose, which was to account in advance for the lost fifty-five minutes. When they arrived at their destination, they still thought it was twenty minutes past eleven, their expected time of arrival had the incident never happened; and they were astonished to find that it was a quarter past twelve. Clearly they had not accepted in their minds the 'explanation' offered by the cover story to account for the lost time, for they had not regarded it as lost; the curious experience of the narrow lane had seemed to them, while interminable, not to occupy time, to be outside time. They remained puzzled by this inexplicable experience, of the strange narrow lane mysteriously interpolated into a familiar journey, but they pushed it aside, as it were, and did not think of it as part of the timed sequence of their journey. To this extent, the psychological device used by the Janos people may be said to have failed.

Nevertheless, it created a diversion which with other persons, in other circumstances, might have succeeded in 'covering up' the visit to the spaceship, if that was the purpose of the Janos people. One wonders how many other close encounters of the fourth kind have been more successfully covered up, so that the people involved have no memory of them. Certainly other, similar incidents must have occurred; it seems unlikely that the incident described in this book is unique.

As to my interpretation, psychologists will no doubt hasten to put me right. The 'experience' of the interminable narrow lane, even taken by itself, apart from the real spaceship experience, should provide plenty of material for controversy among experts in mental science. To such experts, I would say one thing: you will find, when you meet the Janos people, that you have met your masters.

CHAPTER TWELVE

The Janos People

WE HAVE NOW completed the straight narrative account of what happened to the English family who spent nearly an hour in a spaceship, and of all that they saw and were told.

For the convenience of readers, at the risk of some repetition, I will now summarise the main items of information about the Janos people that have come to light during my investigation of this incident. The facts are all drawn directly, and without alteration or embroidery, from the statements of the members of that family group; mostly I have worked with John, his sister Frances, and John's elder daughter Natasha.

These statements were, in many cases, brought to light in sessions of hypnotic regression, used primarily as a means of releasing the amnesia which was deliberately induced by the ship people, to protect the family from the embarrassments of premature publicity of the wrong kind.

As many people will know, hypnotic regression has the further advantage that it enables the subject to re-experience the event, several times if necessary, to give opportunities to look more closely at the scene, and recover details that in ordinary memory would be lost or less sharply defined. Much of the detailed information, however, came out of the subsequent 'follow-up' sessions, in which the hypnotist was not present; and some very important items have just 'popped up' into consciousness at odd times, not in a formal session at all.

This is not the place for an essay on regressive hypnosis: I think it may be helpful to other investigators to add simply that, provided (and it is an important proviso) the hypnosis is controlled by a highly skilled and very experienced professional hypnotist, with a constructive and responsible attitude to the case, it can be most valuable as an aid to the

The Janos People

elucidation of close-encounter cases, especially, as is common in such cases, where some degree of amnesia is involved.

The following is a condensed account of the main facts about the Janos culture, within the limits of what we have learned; I have provided also some general background information where I think it may be helpful.

The Planet Janos

Janos, before the double disaster which rendered it uninhabitable, was an Earth-type planet revolving about a star (their sun) which is *"several thousand light-years"* distant from us. For the general reader this places it well inside our own Galaxy, which has a diameter of about 80,000 light-years; indeed in galactic terms it is a comparatively near neighbour.

'Several thousand light years' (Uxiaulia's phrase; the Janos people seem to prefer not to give exact figures) is nevertheless an enormous, almost unimaginable distance: *"further away than you have ever dreamed of"* is, again, Uxiaulia's somewhat poetical expression.

A light-year, as I am sure most readers know in this space-conscious age, is a measure, not of time, but of distance: it is how far a light beam, from a star for example, will travel in the vacuum of space, in a standard year – that is, 5,886,000,000,000,000 miles or 9,470,000,000,000,000 kilometres.

Light reaches us from the Moon in one and one-third seconds; from the Sun in eight minutes; from the nearest star (Proxima Centauri) in four and a quarter years; from Sirius in eight and a half years: so we say Sirius is eight and a half light-years away. Distances of other well-known stars are: Vega 26, Arcturus 41, Capella 47 and the giant Betelgeux 600 light-years.

So several thousand light-years, whatever may be its precise value, is what one might call in the vernacular 'a heck of a long way'. The Galaxy (our galaxy – there are millions of them) has, in common with many others, a spiral structure caused by its own slow rotation, rather suggestive of a giant Catherine-wheel firework.

There is a series of 'arms' spiralling out from the centre. Janos's sun is most probably in the same part of the same arm that our Sun is in; just possibly it could be in the nearest part of the next arm. (Just to cut us down to size, the nearest galaxy to ours, that named after the constellation Andromeda, and very similar to our own galaxy, is two million light-years away; it is the furthest object that can be seen from Earth with the unaided eye.)

Janos cannot be very different from the Earth in size or gravity; the spaceship people appear comfortable in normal Earth gravity, and (though they are perhaps slightly slimmer) their body structure and general build scarcely differs from that of normal Earth people. Our visitors to the spaceship found normal Earth gravity maintained on board, though it is clear that they could control or vary it if they found it more convenient to do so.

Most probably the sun of Janos is closely similar to our Sun; there is nothing to suggest an important difference. Our Sun is, after all, a star of a very common type. The spaceship people do, on the whole, tend to operate on the Earth's surface at night; but this is more probably to escape discovery than to avoid the sun.

Janos itself, as seen from space before rockfall, had a general greenish-brown tint, varied by patches of blue water. Cloud was not seen, as it would be in any of our pictures of the Earth; but their pictures of the Earth do not show cloud patterns either: so possibly they are using a technique which does not show them – this would be useful in surveying a lot of planets for possible settlement.

There were large areas of open water, seas or oceans, and a great many lakes. Nevertheless, the total area covered by water was said by John to be less than the land area, possibly a third of the planet's surface; in this, Janos differs from Earth, which has more water than land. There were considerable tracts of dark green which to John suggested forests, as well as varied areas which he thought represented cultivated land. Unfortunately, we have no orbital pictures of pre-disaster Janos except those taken from quite a high orbit; all the closer views are of the planetary surface after rockfall.

There are (or were) two moons, both small compared with

ours; the inner moon, called Saton (pronounced Záton) was too close for stability, and its break-up caused the disastrous rockfall. The outer moon, which we think is called Sarnia (pronounced Zárnia) is smaller than Saton.

Climate and Vegetation

"On Janos it was always warm" – according to space-pilot Uxiaulia. We have only a few outdoor pre-rockfall scenes of normal life, but the clothing in these suggests a summer's day in England; it certainly does not indicate any extreme which we would find uncomfortable. Of course, we must remember that the Janos people may not have occupied the whole planet; their numbers were to be reckoned in millions rather than in thousands of millions like ours, and they could have kept to the pleasanter parts.

If we are to understand from Uxiaulia's statement that there were no pronounced seasons like our summer and winter, this would imply that the planet's axis of rotation was close to being perpendicular to the plane in which it revolved round its sun – what we would call the plane of the ecliptic. It would also imply that the orbit of Janos around its sun was not far off circular. Seasonal changes on the Earth arise from two causes: one is the inclination, by an angle of 23½ degrees of arc, of the axis of rotation from a perpendicular with the plane of the ecliptic; the other is a moderate eccentricity of the Earth's orbit round the Sun, which brings it nearer to the Sun in December and takes it further away in June, so that the northern winters are moderated and the southern Antarctic winters intensified.

The mild summery climate also suggests that the amount of heat received by Janos from its sun is not very different from that which on average the Earth receives. This could mean that the heat given out by the sun of Janos is similar to that of our own Sun, in which case the distance of Janos from its sun would be much the same as Earth's; or a comparable balance could be struck if, for example, Janos were a little further away from a slightly hotter star, or nearer to a cooler one. But it is likely that the conditions matched closely those familiar to us.

Janos vegetation differs from ours in one important respect – the leaf colour is a deep bottle green; the delicate lighter greens familiar to us were not seen by our witnesses in films shown to them.

The 'trees', also, differ from those of Earth in that, at least in the examples seen, there is no trunk or bole; several large branches come together directly from the ground, spreading out in the way our tree branches do. On one occasion, a spray of foliage was seen close to the camera; the leaves are described as oval with a slight point, with veins and a midrib similar to ours, but distinctly heavy and fleshy. The witness compared the leaves to those of a rubber plant.

Several shrubs or bushes were seen, with the usual dark bottle-green foliage, but bearing also what appeared to be flowers, described as large, rather like a big rose or a peony, and (in one example seen closely) of two colours – some red and some pink.

Some 'trees' by the lakeside bore large oval fruits of a mustard-yellow colour, not unlike a melon.

Buildings and Transport

We have very few pictures of buildings; the only good clear ones are of houses. Frances had a fleeting impression, during the rockfall scenes, of larger buildings in an urban setting, crashing to the ground; the only buildings seen in post-rockfall pictures were totally ruined – "just bits of wall standing, like a bombed city".

The clearest picture was that shown to Frances by Uxiaulia – a still photograph shown on a screen – of his own home before rockfall, with his wife Vurna and their two young children in the garden in front of the house. The plan was rectangular, with a simple pitched roof without chimneys. The roof was covered with large square tiles of a mid-grey colour, which did not overlap, but lay in one plane. The end gables were embellished with a decorated bargeboard and a carved finial.

The house was of one storey only; "*we do not have an upstairs like you*". The white walls were constructed of horizontal planks, jointed together, of a material which to Frances suggested wood.

There was a door opening in the centre of the front wall: its shape was exactly like those seen in the spaceship, with rounded corners and a slightly arched lintel; so my original assumption, that the spaceship doors were designed that way for structural reasons, may not be correct. On the other hand, the reader will soon understand that there is a reason for thinking that the house door openings on Janos could have been copied from spaceship doors.

Occupying the greater part of the wall to either side of the doorway was a large window; this was not built up from panes fitted into a frame, but was formed of a single very large continuous sheet of glass (or other transparent material) bowed out into a segment of a circle, like an old-fashioned bow window in function, but having a more 'modern' look. Frances could not see into the interior of the house at all.

In another picture, the camera was looking across a valley to a hill opposite; the valley was filled with a number of houses similar to the one already described, which also spread up the slope of the opposite hill. There was vegetation around the settlement, but not noticeably among the houses. The visual impact of the picture was made by the pattern of roofs; the general impression was of a village or residential suburban scene.

There was nothing to indicate access roads, and probably none were needed. Uxiaulia told Frances: *"Our transport is different from yours; our cars float above the ground"*. Private cars could thus 'float' among the suburban or village houses, or to a storage place; a local road system would not be needed, though the ground would not have to be unduly obstructed.

John certainly saw, and – through the medium of a film camera lens – actually rode in such a float-car; though this one was bigger than a private car would need to be. He describes how it glided over the rough ground, never touching the surface, and how it continued to hover, a few feet up in the air, even when it came to a temporary standstill.

This vehicle was an elongated oval in shape, with a pair of large cylindrical projections at the rear end, parallel with each other and with the axis of the vehicle; they could have been

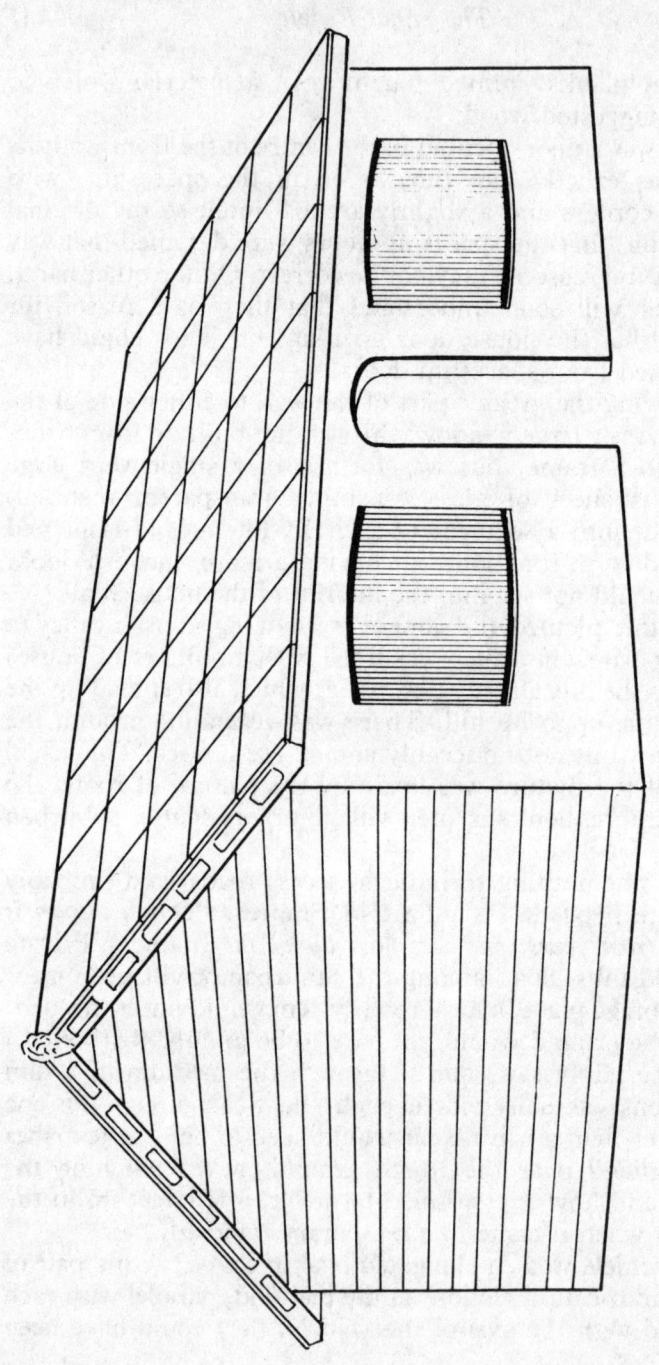

18 A private home on the planet Janos

engines. Internally there was a windscreen giving a view of a curving 'bonnet' in front; the vehicle carried headlights which were used in a tunnel. The body of the vehicle was a dark gleaming metallic colour.

In December 1979, John told me that he had recently had a dream, in which he found himself back in the spaceship, among the silver-clad crew. He was standing, with Anouxia beside him, looking into a large vertical screen let into the wall; this was not the horizontal screen, forming part of the navigating table, in which he had earlier seen pictures. Three or four of the ship people also stood around, watching the picture. John could not say where this was in the ship.

In the screen, he saw, to begin with, a country view, with a lake in the background; there was a boat on the lake, in movement, but it was too far to make out detail. The daylight seemed to be not very bright.

He turned left (that is, the camera did so) and faced up a rather steep natural grassy bank, perhaps twelve feet high. He went up the slope; John said that it was not too steep – it was not much effort to climb up (this is typical of the subjective feel of actually taking part in the scene that Janos films give).

At the top of the bank, he came out on to a road; he looked first to the left, then to the right, where the road, straight as an arrow, sloped gently down and swept away into the dim distance.

The road itself, he said at first, was like a motorway; it was very wide and completely smooth, surfaced with a dark-coloured material. There were no lane markings, no central reservation, no hard shoulder, no signs. Afterwards he said it was more like a runway at an airport.

On either side of the road were houses; there were perhaps a hundred of them, generally similar to the one Frances was shown. He also remarked on the pattern made by the roofs; perhaps this is something the Janos people admire, and may have commented on. One thing John does remember is that Anouxia, when they were looking at the houses, said: "*Some of the buildings are like your shops*".

The nearest houses were quite near, perhaps fifty feet or so; they were below the level of the roadway, so that the road verge was about on a level with the roof eaves. Further along,

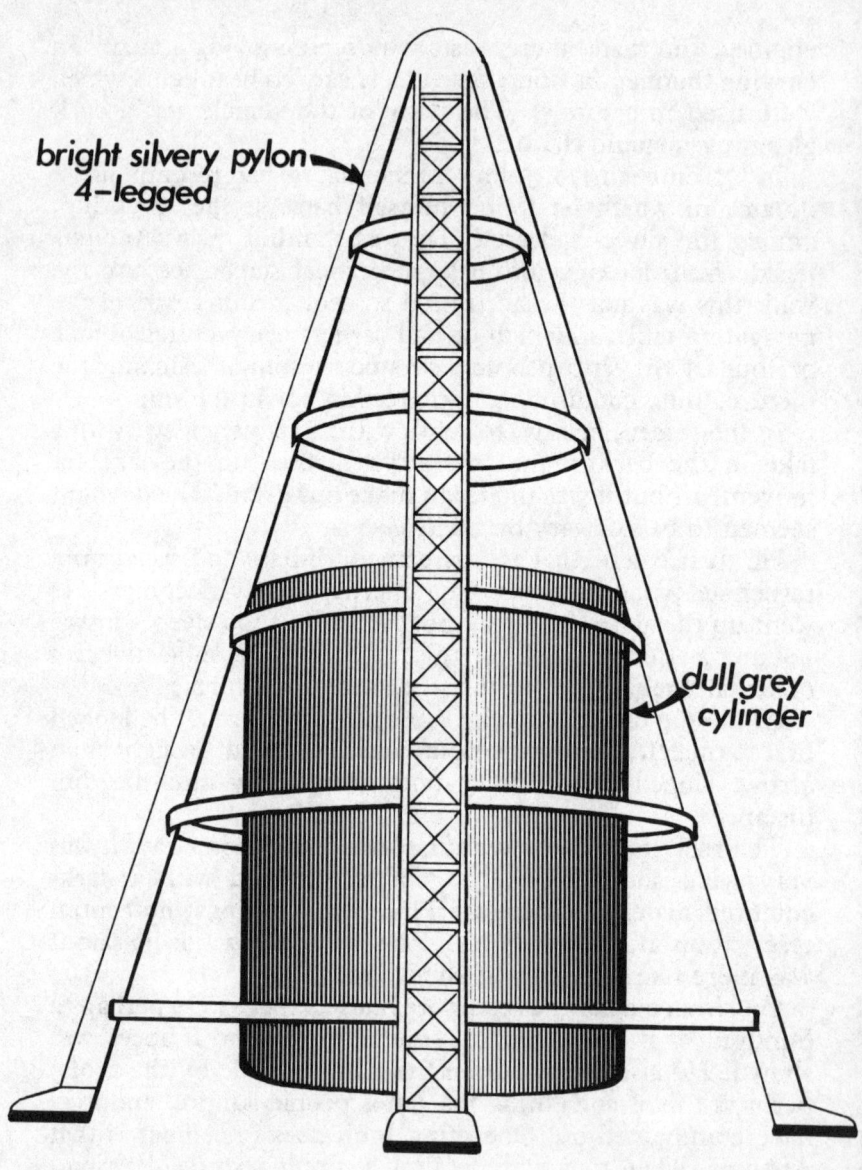

19 A nuclear power station on the planet Janos

as the road sloped gradually down, it came down more towards the level of the houses.

John said there were no branch roads or turnings; there was no way you could drive off the main road into the group of houses. But with float-cars this would not be necessary; the main trunk road would provide a free unencumbered way for fast-moving long-distance traffic, still off the surface; but local access and delivery could be floated over rough ground.

He saw no traffic on the highway, and no people; when he looked out again over the countryside, by the lake, it seemed rather dark; but along the roadway and in the built-up area, it was brighter, though he saw no lamp standards.

In our planet, many years ago, there was some talk of ionising the air above roads and towns, using the principle of the aurora borealis, to provide a soft, even illumination which would need no visible fittings; I have wondered, since hearing this latest report of John's, whether Janos used such a system of lighting.

Power and Industry

Electrical power was generated by nuclear power stations, different in appearance from ours; an example seen by Frances was aptly described by her as "like a gasometer inside the Eiffel Tower". There was a huge vertical cylinder of a dull grey material, supported only at four points of its upper circumference by a four-legged tapering pylon of bright shiny metal, composed of lattice girders with a criss-cross pattern, joined by a series of rings. The cylinder was suspended by the pylon in such a way that its lower end did not quite touch the ground, though it must have been very heavy. The pylon came together at its apex into a smoothly rounded cap.

The scale was difficult to assess, for lack of anything to compare it with; Frances had the impression, however, that it was of very great size. She is almost certain that the fuel used was uranium; she does not remember that it was so named, but that was how she identified it in her mind. In view of what happened, it seems very likely to have been uranium.

Of the industrial achievements of Janos, we can judge only

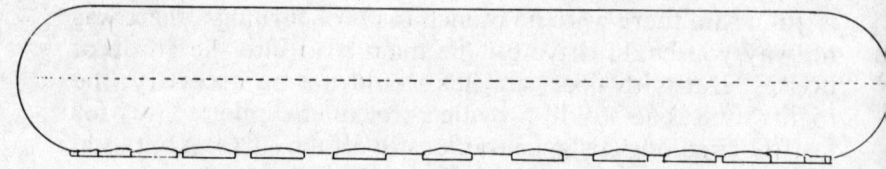

The Janos 'flagship' in edge view, as shown to Frances

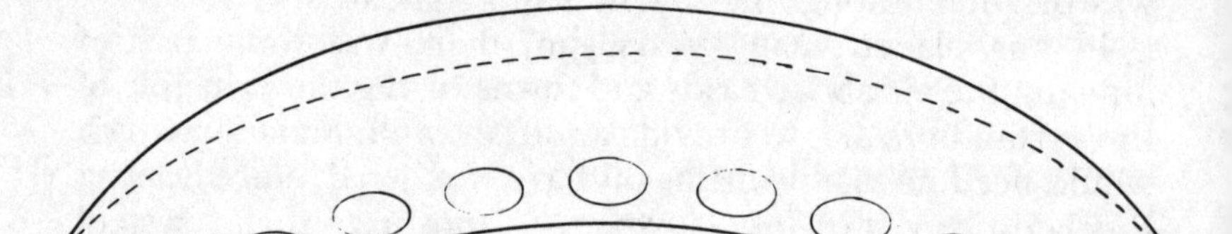

entry ports for spaceships

20 Probable appearance of 'flagship' in oblique view : the overall diameter can hardly be less than three miles (5Km)

by inference. However, it is impossible not to be deeply impressed by the sheer magnitude of the task of building a space fleet to carry, over such a vast distance, the entire population of the planet, numbering many millions. How long this task required, we do not know. We have been told that the average working day on Janos was much shorter than on Earth; and that most of the work was done by machines and computers.

The creation of just one spaceship such as the one visited – regarded by the Janos people as a 'small' ship – argues an enormously advanced and complex industrial organisation; and its dependence upon electrical energy no doubt reflects the electricity-based energy-pattern of Janos industry. There are many such ships, some of them of great size: one in particular, which I have called the 'flagship' of the fleet, was shown to Frances by a technique which I have described below under Telepathic Communication.

It is of really stupendous size, and appeared to be in the form of a ring; but because Frances saw the ring edge on, she could not tell whether it was filled in as a disc, or whether it was just an open annulus. Set around the outer rim, in a circular pattern, were many huge entry ports for spaceships to come and go.

Despite its high technology, there is much that is familiar about Janos spaceships, as exemplified by the one visited. There are bolts, for example, though with octagonal boltheads, not hexagonal. There are voltmeters. There are things that strongly suggest transformers. There are video screens, for displaying technical data as well as for showing films and for monitoring external cameras. There are powered ramps which move forward automatically when a person steps on them.

Somewhere out of our sight, in the spaceship, there must be workshops, stores, drawing offices, laboratories, computer rooms, test benches, repair and maintenance equipment, as well as rooms for living and social life. On the planet, on a vastly greater scale, the same things must have been there; and we can assume that the great ships of the fleet carry with them the entire range of industrial equipment of Janos, as well as its population.

Food and Animals

Our information on these subjects is scanty. "*We have animals for food; we do eat some meat, but mostly we eat the things that grow*": this is almost the sum total of what we were told verbally. Yes, one other thing; at the lakeside barbecue they were cooking and eating pieces of dark-coloured flesh; Uxiaulia called it "*meat*" and said "*we get them from the rivers*".

Frances also saw that they were eating the mustard-yellow melon-like fruits from the 'trees'; they seemed to be popular.

We do not know whether the Janos people practise agriculture in our sense of the term: "*we have animals for food*" certainly suggests that the animals are kept for the purpose.

We were told that the Janos people do not keep animals indoors as pets; but the presence on Janos of at least one dog or wolf makes it likely that, at least in the past, they used primitive wolf-dogs as hunting dogs, or conceivably in the way we use sheepdogs.

The 'meat' caught in the rivers suggests fisheries; but this is to use the term in a loose sense – they do not have to be fishes as we have them on Earth.

On a picnic occasion, the people ate with their fingers; but indoors they may have been more formal – we do not know, and Frances was shown no picture, unfortunately, of a domestic interior.

Clothing and Hair Styles

Information about clothing, on the other hand, is very detailed. A fuller account may be found in Chapter 6; I will summarise here.

Four distinct kinds of clothing have been described: one in the spaceship and three on the planet. In the spaceship, the fifty-odd crew members seen were all in uniform, which was the same for men and women. Basically, this consisted of a one-piece close-fitting garment covering the whole body, in a fabric faced with gleaming metallic silver, but quite supple and flexible. Our witnesses were of the opinion that the metallic finish was actually silver, not merely a silver-looking metal.

Most wore a belt of the same material, transversely ribbed in an example seen close; to the middle front of the belt was secured a circular badge of about three inches diameter, the details of which are given below under Flags, Badges and Insignia. In one example seen close, the badge was secured to the belt by an elaborate silver clasp.

The two senior officers, Anouxia and Uxiaulia, did not wear a belt; instead of the badge they both wore a large, absolutely plain white disc on the chest, about five and a half inches diameter. The witnesses were not told the significance of these discs.

In the spaceship, most of the crew of both sexes wore, as part of the uniform, a close-fitting silver 'balaclava' helmet, showing only the face. Some, at least, of the helmets had an ear-covering, in very thin silver, which was modelled to follow the lobes of the ears. At least one man, Anouxia, wore gloves of thin silver.

Frances saw, at very close range in one example, a mark along the left shoulder line which she interpreted as being similar to a concealed zip fastener. (Observers of spaceship uniforms have often wondered how the wearers get into and out of them, since they seem to have no fastenings.)

The shoes were of black or grey uppers, with very thick white spongy soles with no heel; they enabled the wearer to cling to the deck in 'zero-gee' conditions. (The absence of a heel is one of the most consistent details of published descriptions of the clothing of flying saucer personnel.)

We were told that the Janos people did not wear silver clothes on the planet; they were for the ships, and are a uniform.

In the spaceship, men who wore no helmet had the hair cut very short, American 'crew-cut' style, brushed straight up in front; the few women in the ship who were bareheaded wore their hair long, page-boy style, brushed out free and curling under at shoulder level. We may, I think, assume that all helmeted people, of either sex, had the hair cut short, though we never at any time saw a bareheaded woman with short hair.

The people who were slowly dying of radiation sickness, on the planet after the power stations blew up, wore a monk's long-skirted habit in black or dark brown, with a deep hood or cowl over the face; this, we were told, was special clothing, to give them some protection against the lethal dust; it was not their normal clothing. Possibly it may have served also to identify them as contaminated by radioactivity.

On the pre-rockfall planet, in normal life, we have examples of clothing in two situations: at home, and at leisure. We did not see people at work, and do not know what they would wear.

Domestic clothing for women and children is seen clearly in the photograph of Uxiaulia's home, which shows his wife Vurna (a woman 23 years old) and her two children, in the garden.

The basic garment for all three is a pair of dungarees over a white jumper; the straps of the dungarees are fastened in the front of each shoulder by a white circular buckle.

Vurna, and the girl aged about five, both wore red dungarees; but the little boy, about three years old, wore pale-blue dungarees.

In both children, the jumper came up to a high round neckline, and was long-sleeved; Vurna's jumper was shorter-sleeved, the sleeves ending above the elbow.

Both children wore white shoes; but the mother's shoes were red.

Vurna's fair hair was brushed out naturally, and curled under at the ends, at about chin level.

The little girl had curly, yellowy-flaxen hair; most of it was free; but on each side of the temple, a bunch of hair was brought out through a red circular hair-grip or slide, so that the two bunches of hair stood out to the sides.

In the film of the lakeside evening barbecue party, clothing was of three kinds. Some of the men, including the one operating the barbecue, wore only a pair of dark-coloured swim-trunks, similar to those familiar to us.

Other men wore an overall suit something like a track-suit, with a broad belt at the waist. In one example, there were white stripes down the outer edges of sleeves, body and legs. In a boat with a mixed crew, both the man and the woman wore a red track-suit; evidently this was considered more convenient for activity than the garments I am about to describe.

The women who strolled or sat on the lake shore with their companions were more fashionably dressed. All wore variations on a common theme: this was basically a long-sleeved bodice with a high round neckline, worn with an almost ground-length full skirt, draped in overlapping folds down the left side. The whole thing was held together by a large round metallic-looking clasp, of an abstract floral design, on the right hip.

The bodice was, with one exception, white, in a filmy material like a chiffon or a fine nylon. The skirt material allowed some variety in colour scheme; but in most the ground colour was white. The skirt material was printed with a large repeating pattern which, although details varied, was basically an abstract floral motif, related to the design of the clasp. Red and pink designs were prominent.

One woman in the foreground wore a black bodice over a white skirt with the floral design printed in black; her hip-clasp was also black. She had a black head-covering of some kind; she was the only woman seen by our witnesses on the pre-rockfall planet who was not bareheaded.

Flags, Badges and Insignia

Light can often be thrown on the nature of a culture, and sometimes on its derivation, by its use of visual symbolism. We are fortunate in having some details of this kind from Janos.

The pennant flags flown by the two speed-boats were both triangular, the length being about twice the hoist. One of the two was fishtailed, possibly to distinguish it from the single-pointed one, though the boats themselves were of different colours.

The ground colour of both pennants was dark blue; a white disc upon it nearly touched the hoist and the two edges. Upon the disc was a device, again in dark blue, consisting of a

abstract floral motif

21 Janos woman's evening gown for a social occasion

looped line, with a round spot where the loop crossed over.

The circular badge worn on the belt by most of the spaceship crew, about three inches overall diameter, was in two concentric zones: an inner circular zone of two inches diameter with a surrounding annular zone half an inch wide. The inner zone was white, not silver; upon it, drawn in black lines slightly raised from the ground surface, was a stylised representation of a 'flying saucer' ship in side view: from the centre of the under side, two lines, diverging downwards, suggested the outline of a beam directed to the ground. Possibly this may have represented the 'survey' function of this class of ship.

The surrounding annular zone was black, with the two circles which marked its limits slightly raised. Upon the black zone was a continuous pattern of straight silver lines set at odd angles to each other; it was not regular enough to be just decoration, and both John and Frances, who also described the badge, were sure (perhaps were told) that it had a meaning, that it was writing. John has attempted to sketch some of it, and I have indicated his impression in the drawing; but this should not be taken as an accurate representation. I would guess that the lettering round the badge is in a stylised inscriptional form, rather than being normal script. It suggests a form of writing which was runic in origin – lines cut into wood with a knife.

John also saw, at one point, a doorway of the usual form, but surrounded by a decorated architrave, about four inches wide; the markings on it were generally similar to those on the badges.

Some people in the spaceship also had special flashes, such as the yellow bands on the shoulders, worn by Serkilias and Cosentia; from the way in which Serkilias touched them when she said 'medical', meaning that was their job, they were probably flashes indicating 'medical personnel'.

One man, at least, the 'big' man who escorted Frances in the corridor, had long tapering white markings on each side of his chest, beginning with a broad end below the shoulder, and tapering down vertically to a point at about heart level. The meaning of this was not explained to us.

Speech and Language

When the Janos people in the spaceship spoke to each other, the visitors were aware that they were speaking a foreign language; they said it was like hearing a foreign station on the radio, or listening to Spanish, which they do not understand, when they were on holiday in Spain. But it did sound like a European language.

When the spaceship people spoke English, it was a standard unaccented speech as spoken by English people of good education; John said, modestly: "They spoke better English than I do". There was no trace of local or regional accent, and no foreign intonation; it was definitely the English of England and not of Scotland, America, Canada or Australia; nor was it English as spoken well by a German, Dutchman or Scandinavian, for example. It was only in the use of idiom that their English speech occasionally betrayed itself as a learned language, rather than a mother tongue.

Of the Janos language, we know very little, but just enough to be interesting. We have, so far, thirteen proper nouns with their pronunciation: JANOS SATON SARNIA ANOUXIA UXIAULIA VURNA AKILIAS SERKILIAS COSENTIA SAUNUS VONASON FAUN and PHUSANTHEAS. The spelling in English letters was provided by the spaceship people for the first five of these; the remainder are our spelling of a spoken word, though Natasha knew that Phusantheas began with a P, and the spelling of the last four may have been given to her by Akilias.

I was at once struck by the affinity of the words, taken as a sample of a language, with archaic Greek; Phusantheas has a very Greek ring, and Saton is actually an old Greek word meaning a corn measure; here it is applied to the inner moon of Janos, the one that broke up. All of the words are possible, considered as words in a language akin to old Greek. Moreover, the thirteen words all hang together linguistically: they all belong to the same kind of language; this is a strong argument against any suggestion that they were invented by the witnesses, who would need the imagination and linguistic knowledge of a Tolkien to have thought them up for themselves.

Telepathic Communication

In addition to normal speech, whether in English or in their own tongue, the Janos people, or at least some of them, have the capacity to transmit visual images, by a form of telepathy, from their minds directly into the mind of another person. These images are sharp and clear, like a film with colour and movement; they are not in the least vague or insubstantial. They look like actual scenes, with the usual Janos quality of pictorial realism.

This technique was employed to show things for which a film was not available: Frances saw the power station and the 'flagship' of the fleet by this means; and both John and Frances and perhaps Natasha were given spellings of words by a succession of single English capital letters, dark on a light ground, shown one after another, rather less than a second apart.

Normal conversation was in ordinary sound speech: John, in a hypnotic re-experience of a speech sequence, checked up by my request on lip movements, and said afterwards that the lip movements of the Janos speaker corresponded to the English words that John was hearing him speak.

Nevertheless, I have a strong impression, from certain things Frances has said in follow-up sessions, that the ordinary sound speech is reinforced by a telepathic communication of the underlying meaning. To some extent, I believe this is often so in normal Earth speech, between persons who know each other well and think alike; but this was sharper and clearer.

To give an example: when I questioned Frances about a phrase used by Uxiaulia – "*together we can conquer all space*" I remarked that it did not seem to go with their professed renunciation of war, she at once corrected me, saying that he had not meant by that, conquest in the military sense, but in the sense of space exploration. Uxiaulia is a space explorer pilot by profession, so this may well be correct; nevertheless it was said in a context of attack and defence, so that my query was a reasonable one.

Personality and Politics

Our family found the Janos people whom they met in the spaceship, as personalities, extraordinarily attractive; they were strongly drawn to them, and even now have a sense of loyalty to them, even of identification with them. After a brief emergency when the ship had to be lifted because someone was coming, Frances said, in tones of relief: "It's all right; they've gone". The reader should understand that she said it in a re-experience of the event through hypnotic regression; she did not necessarily say it at the time, though it was probably in her mind at the time.

John, asked to sum up his impression of the Janos people, said: "They were so friendly; there was no hostility".

Our combined impression of the Janos people is that, while there are clear indications of differences of individual temperament and point of view, they are very united, and very close to each other; doubtless their common tragic experience has bound them together; but they seem by nature very united.

They clearly attach great importance to everything being done by general agreement; there is no place in Janos society, as we understand it, for the dictator, king or boss-type. In the spaceship, of necessity, some give orders and others obey; but our witnesses sensed an underlying equality.

Concerning what I have called the 'flagship' of the fleet, Frances was told that it was where all reports go, and where all the big meetings are held.

Physical Type and Race

All the people seen in the spaceship, and all those seen in films of the planet, were without exception of the Nordic European type – very fair-skinned, with light blue eyes and yellow-blond hair. The men averaged around six feet in height, and were of slim build. The women were smaller, about five foot four to five foot six, and of slight build, with a light slender body, small breasts and slim hips; it is possible that a majority of the women seen were quite young, and this is the impression they made on our witnesses, though they may not have been as young as they appeared, quite apart from the relativistic time-shift. Serkilias, when John happened to see her very near, seemed to him "most attractive". The women themselves evidently felt that their eyes needed more emphasis, for many of them used make-up to darken the naturally pale eyelashes and make the eyes seem larger.

We cannot, of course, be sure that all the Janos people conform to this type; our witnesses saw less than a hundred people in all. There is a possibility that other racial types may be represented in other ships; if this is so – and it does seem to be suggested by some incidents reported from other parts of the world – it must mean that, whatever the circumstances in which their ancestors left Earth long ago, more than one local group was involved.

However, a large proportion of reports of normal spaceship people, as distinct from the various dwarf and goblin types described, do conform to the fair blue-eyed type described here; and it may well represent the typical Janos person.

It is remarkable, on any theory of Janos origin from Earthfolk ancestors, that the so-called 'black' or 'coloured' races of mankind appear to be totally absent from reports of 'space' people, so far as my own reading goes. This circumstance, in itself, lends weight to the idea that the Janos people originated, very long ago, from a comparatively small area of Europe, in an age before the various races of mankind began to mix as they have done in modern times.

CHAPTER THIRTEEN

Under Whose Flag?

THE DEEPER WE have dug into this matter of the Janos people, the clearer it has become that they are, in origin, Earthfolk who have been away from their planet of origin for a very long time. They themselves tell us:

"*We knew about Earth from the past . . .*"
"*We knew where it was in the sky . . .*"
"*Looking at you people and the planet, it is like stepping back into one of our own history books; to us, you are living history.*"
"*You are our people, because you are the same as us.*"

The Janos people say that, although they explored many planets within reach of their ships, to survey them as possible homes, against the time which they knew would come, when they had to leave Janos for ever, they never came to Earth, because it was too far. The recent arrival in our solar system of the great migration fleet is their first visit since their ancestors left the Earth, in the remote past.

At first sight, this statement appears to create a problem of timing. Flying saucers have been with us for a long time: Livy and other Roman historians refer to the *clipei volantes* or flying shields, which the Romans regarded as portents; if you take two Roman shields and place them edge to edge, you will have a very fair model of the shape of a typical flying saucer. It is with this analogy in mind that Italians use the word 'clipeology' where we would use 'ufology', for the study of flying saucers and related phenomena.

Not all UFOs are of the biconvex lenticular shape, with its characteristic dome or blister centrally in the upper and lower surfaces; but it is by far the most frequently reported. The

Janos ship visited by our family, described in this book, is of this kind.

Reports of unearthly flying machines, fiery chariots and what have you, go right back through history, as far back as ancient Egypt and Babylon. They turn up in the Old Testament, and in the ancient Sanskrit. Are these, then, not from Janos?

It is hard to see how they could be: if we accept, as we must, the statement by Uxiaulia that they aged by *"two of your years"* during a journey of *"several thousand light-years"*, it follows that their transit velocity was very close to the speed of light; thus the time taken, reckoned by planet time, cannot have been much more than light would take to travel the same distance, and it certainly cannot have been less. In other words, the event of the rockfall on Janos, which precipitated their departure, happened several thousand years ago, by our time.

Whatever figure we attach to 'several', then several thousand years ago, by Earth time, the fleet left Janos on its way to Earth. Nothing can reduce that figure substantially, reckoning by Earth time. If some ships travelled a little faster, a little closer to the speed of light, this would reduce the time the journey would appear to them to take, by ship's time; but I would not think that it could get them here substantially sooner by Earth time – not to make a difference of thousands, or even hundreds of years – unless there is something about the peculiar mathematics of close-to-light velocities that I have missed, being a mere biologist. I am open to instruction.

I am inclined to think, therefore, that reports of UFOs before modern times, stretching back into early history, do not relate to the Janos story, but are reports of vessels coming from other planets. The Roman references to 'flying shields' are certainly suggestive of flying saucers; but of course we are not in a position to say that only the Janos people ever had flying saucers; they could have learnt the design from someone else. Although we do not fully understand the mechanics of saucer flight, it is apparent from what we have learnt from the incident described in this book, that the circular, lens-shaped form of spaceship arises naturally from the need to accommodate the rotor mechanism, which dictates a circular shape.

Any civilisation employing the rotor mechanism as a factor in space flight would necessarily use flying saucers, for any ship intended for planetary landing and lift-off; it appears, from what John saw and was told, that the rotor spins in order to obtain annulment of gravity during lift-off: "*If we spin* (the rotor) *fast enough, there is no gravity to hold us down to the Earth*". In space flight, the rotor is still.

The 'big' ships, built in orbit, which can never land on a planet, need not be circular, lens-shaped; they could, for example, be cylindrical – very large cylindrical spaceships have occasionally been reported, of which the most famous example was the one which exploded (a nuclear explosion) two miles above the ground in the Tunguska region of Siberia, in 1908 – see *The Fire Came By*, by John Baxter and Thomas Atkins, for a full account of this incident, which for many years was mistakenly thought to have been the fall of a giant meteorite.

But there is, of course, nothing to suggest that the Tunguska spaceship came from Janos: its use of nuclear propulsion power – the ground below is radioactive to this day – might seem a thing the Janos engineers would avoid; but one must remember that the Janos big ships were all designed, built and in use before the nuclear catastrophe occurred, so we cannot, on those grounds alone, rule out the possibility that it was a Janos ship; whatever its origin, it arrived in 1908, but got into trouble in our atmosphere; eye-witnesses say it was on fire long before it exploded; also that it twice changed course, showing that it was still under control.

One spaceship visit in our history – described in *The Sirius Mystery* by Robert Temple – was clearly not from Janos. It is specifically associated with a planet of the Sirius system, only eight and a half light-years away – in stellar distances a close neighbour, compared with the 'several thousand' light-years of Janos. Also it is very much too early; the landing tradition is given in ancient Babylonian and Egyptian texts. Babylonian picture-seals illustrate the visitors, representing them (no doubt a simplification) as half-man, half-fish. The Babylonian text says that they "taught mankind the arts of civilisation;

Under Whose Flag?

and having remained for a while, they departed whence they came".

Another early tradition, from India, given in the ancient Sanskrit texts, also describes, most graphically, the arrival of spaceships bringing people of an advanced civilisation, who remained for many years, establishing cities; but they also eventually departed. The Sanskrit account, however, represents them as warlike and quarrelsome, with terrible weapons – the text reads like something out of modern space-television of the more juvenile kind – which does not sound like Janos. Again it is very much too early; Janos ships could not possibly have got here so soon.

What all this is building up to is a general impression that there are other space-travelling civilisations about, not only Janos; and that this has been so for a long time.

* * *

The earliest date of any report which seems to tie up with Janos is 1888. This is the year of an entry in a journal kept by a white man who lived among the Indians in the Rocky Mountains, knew them well, spoke their language, and was initiated into the tribe – the story is told by Brad Steiger in *Mysteries of Time and Space*.

The record is in a private diary, which did not come to light for the best part of a century, when it came into the hands of its author's grandson.

Some readers may be familiar with the sasquatch, momo, bigfoot or 'crazy-bear', an old Indian tradition which in modern times has been revived by many sightings by civilised people, often loosely associated with UFO sightings.

The 1888 diary entry is the only source which appears to link the sasquatch tradition specifically with the Janos story. The link is made in the reported Indian description of the men in the flying saucer – what the Indians call a 'small moon' – as having very short hair and shiny (silver?) clothes.

The author of the diary tells how he was shown a 'crazy-bear' – a powerfully built naked man covered with smooth dark hair – living in a hut. The Indians told him that, from time to time, a 'small moon' came down from the stars 'like a swooping eagle' and landed; the short-haired men in shiny clothes then pushed one or more 'crazy-bears' out of the craft, waved in a friendly manner to the watching Indians (a recognisable Janos touch, this) and flew back to the stars, leaving in the minds of the Indians a feeling – perhaps hypnotically induced – that the 'crazy-bears' would bring them powerful medicine, and that they should feed, cherish and protect them.

I have referred, near the end of Chapter 3, to Natasha's report that Akilias showed her several films, including one of big, powerfully-built naked people, covered all over with smooth dark hair, except around the eyes and on the hands and feet. They were seen first out of doors, on the banks of a muddy stream; then living underground in a cave or burrow, as a family group, with rough mattresses for beds, but no other property.

I have little doubt, bearing in mind the 1888 diary link between 'crazy-bears' and people who certainly looked and behaved like Janos people, that what Natasha saw was a film of 'crazy-bears' or sasquatches living on Janos or, less probably, on a planet near to Janos, but where the Janos people knew and visited them. Natasha's only variation from the 1888 diary description is that she says they had green eyes; although the diary does not mention the eye-colour, two of the more recent published reports of sasquatch sightings, the only two to give an eye colour, describe the eyes as green.

Natasha also said that the hands and feet appeared "whitish"; this may perhaps have been an effect of dried mud, for they were in and out of the stream; one of the men stooped down and drank from his cupped hands.

I have no doubt at all that the many reported sightings of the beings variously known as momo, bigfoot, sasquatch or crazy-bear – all Indian terms for the same kind of creature – are of surviving prehistoric men or near-men. Reading the more sober and less sensationalised descriptions, one has a consistent impression of a powerful, muscular build, height about that of a big normal man, but broader-chested, with long, powerful arms. The females, as described, have broad, massive bodies with no waist, and have heavy, pendulous breasts. They have a strong odour, and a loud cry or roar. All

Under Whose Flag?

are described as naked, and covered all over with short, smooth hair, dark or brown, except for the palms and soles and around the eyes. They avoid contact with normal humanity, but can be aggressive. They are flesh-eaters, and live and hunt like a bear.

I see no reason to doubt the 1888 diary, which was private, not intended for publication, by a man living close to unsophisticated Indians and sharing their life.

There could, of course, be living examples of prehistoric near-human types, still surviving in dense jungle or remote mountain forest; in fact there are many reports, some of them by responsible persons, which suggest that this is so. There is no reason to assume that they have ever been away from the Earth, and probably most of them have not. The Himalayan yeti or 'abominable snowman' is almost certainly one of these; likewise certain reports of semi-aquatic 'wild' people from the jungles of Burma.

But one group of 'crazy-bears' do form part of the Janos story, if we accept the coincidence of the 1888 diary and Natasha's evidence. The question is: how did they get to a planet thousands of light-years away?

The answer, beyond any reasonable doubt, is that they went by the same means which conveyed the ancestors of the Janos people – members of our own human species – and also a primitive type of northern wolf-dog, probably accompanying the blond blue-eyed humans; although Janos people do not now keep indoor pets, this does not rule out the possibility that their remote ancestors had the help of dogs in hunting. Frances said the wolf-dog she saw in the film did not seem to belong to anyone; it had the look of a wild animal. Again, hunting-dogs could have gone feral in the new world.

* * *

What I am working round towards is a growing doubt about the circumstances in which the ancestors of the Janos people originally left Earth.

There are two, and only two ways in which they could have gone from Earth to Janos. One is under their own steam, so to speak: by their own technological resources; the other is that they were taken – voluntarily or otherwise – by someone else, not necessarily human. Both alternatives present formidable difficulties; but one of them happened, difficulties notwithstanding.

A long interstellar journey at very high velocity represents a level of technological achievement far beyond anything we have at present on Earth, or are likely to have for some time to come. The 'own-steam' alternative implies that there was such a level of technological achievement on Earth in the remote past. It implies a scientific civilisation far surpassing our own, which has vanished without trace.

It is hard to believe that a civilisation so far advanced would have disappeared totally; though the obliterating effects of ice and flood should not be under-estimated, if such a civilisation were confined to a northern area in which it might suffer severely from a late glacial advance.

Of course, the legend of a vanished high civilisation in the remote past is one of the most persistent of human notions; it goes by many names, of which Atlantis is the most familiar; though the real Atlantis, if it ever existed, was probably much nearer to our own time than what I have in mind. The reader who wishes to go more deeply into this fascinating field will find more material in Charles Berlitz' *Mysteries from Forgotten Worlds*.

But, as far as the Janos saga is concerned, there is a difficulty about the view that they went to Janos by their own unaided efforts. Uxiaulia told Frances that several hundred years before the rockfall drove them from their home-world, they had made their first space-flight to Saton, their nearest neighbour. In fact the whole trouble was that it was too near. He told her that they went there, partly out of scientific curiosity, but partly also because they were running short of certain minerals, necessary for their industry – she is almost certain that he meant uranium – and hoped to find new supplies in their nearer moon.

Uxiaulia gave Frances the impression that they were then new to the business of space exploration, and inexperienced; so that they made mistakes. The danger from Saton was perhaps at that time not fully appreciated; and the Janos people were a little heavy-handed with their mining operations. However, Uxiaulia emphasised that what had seriously damaged Saton was the excessive power of the blast-off from the moon's surface, each time a rocket-ship returned to Janos.

In those days, they were using primitive thermal rockets; and the heat and power of the exhaust jets caused deep fissuring of the crust of Saton, loosening the whole structure, so that it had much less cohesion than anyone would expect it to have. It was little more than a heap of rocks, in a round shape. It was this error which caused the Janos scientists to miscalculate slightly the date of the beginning of rockfall.

When, centuries later, the distance between Saton and Janos was so much reduced as to cause instability, the rocks simply came apart.

Between any planet and any moon of that planet, there is a balancing point along the line drawn between them, at which planet-gravity equals moon-gravity. Normally, as in the case of the Earth–Moon system, this balancing point lies between the planet and the moon, though much closer to the moon.

But if the moon approaches close enough, a time will come when the balancing point lies actually within the body of the moon. When that happens, any part of the moon's crust which lies above that point – that is, nearer to the planet – will be subject to a negative gravitational field (negative in moon terms, that is), and any loose rocks on its surface will fall upwards towards the planet, slowly at first, but accelerating until eventually they hit the planetary surface with tremendous force (though not with the really high velocity of a meteorite).

Once Saton began to fragment, it would swing erratically in its orbit, until it had completely broken up.

Thus it was an early mistake on the part of inexperienced space engineers, using at first far more jet-power for blast-off from Saton than was necessary. Later this was corrected, and just sufficient power only was used; but by then the damage had been done, which caused the rockfall to begin a little earlier than had been calculated.

Saton would have broken up and fallen as a shower of rocks anyway, whether its structure had been loosened or not; the deep fissuring affected only the timing: but this error of timing cost millions of lives. Had the calculations taken fully into account the loss of cohesion of Saton, due to the excessively powerful jet-thrusts of the earliest rocket-ships, the Janos people would all have got away safely in the fleet, without loss of life; and Uxiaulia's young wife Vurna, and their two children, would be alive today.

Earth's own Moon will one day break up and fall as a monstrous rain of rocks; but relax: this will happen so many thousands of millions of years into the future that the Earth will by then be uninhabitable anyway, because of changes in the Sun. If at that remote future date, our descendants are still around, it will be up to them to emigrate in giant spaceships to another planet in another solar system, as the Janos people have had to do.

I have emphasised the inexperience in space engineering of the Janos people when they first went to Saton and caused deep damage to its structure, because it has a bearing on the question: did the ancestors of the Janos people leave Earth under their own steam, or did they go as passengers – voluntary or involuntary – in someone else's ships?

A really long interstellar flight at a speed close to that of light, such as that from Earth to Janos, represents an extremely advanced mastery of space engineering skills. It seems unlikely that the Janos people, if they had such superlative technology when they left the Earth, would have forgotten it all when they got to Janos, and had to develop space engineering all over again, beginning with little hops to the nearer moon (less than our own present level).

This consideration lends weight to the alternative that they went as passengers. This is a disturbing thought; for it implies the existence in the Galaxy of other, far more advanced civilisations, which may not be human.

The fact that the Janos people, as seen in the ship and in films on the planet, are closely similar to each other – though by no means identical; the visitors, though John did remark once that "they all look alike", had in fact not much difficulty in recognising individuals – suggests that at least all of those seen by our witnesses originated from one quite small locality on Earth, certainly in Europe and probably in Eastern Europe; the original group could have been quite small in numbers, maybe no more than a dozen or so as a practical minimum.

Certainly they took dogs with them, of a primitive northern wolf-dog type; on Janos this type has not been since modified by selective breeding, but remains in its primaeval form, which argues that the canine population early went feral.

Conceivably (but why?) they could also have taken with them a group of more primitive near-men, culturally much less advanced than the blond, blue-eyed humans who became the Janos people, and perhaps incapable of civilisation.

All this adds up to the passenger theory being more probable. One would expect the level of civilisation needed to provide the technical means of really advanced interstellar spaceflight to be much more broadly based, involving the whole planet; this kind of thing would not be a parochial, village enterprise. It would seem unlikely, in such a case, that the human species would be represented, as seen in one returning ship and its film records, only by one closely-defined local race-type – the tall, slim-built blue-eyed blond type which we think of as Nordic, though what little we know of their language suggests an affinity with archaic Greek.

I suppose it is just possible that the Janos colony, after landing, suffered a total loss of civilisation, and had to learn everything all over again; but it does seem unlikely, in view of their quite definite and precise information about Earth, including the technical data – very complicated of necessity – required for navigation back to the Solar System. Such information could scarcely have survived a major breakdown of civilisation. Also, there has not really been time for such a regenerative process, without putting their original departure from Earth improbably far back into the past.

I say 'improbably', because the Janos people are present-day human in type. If their departure from Earth had been much more than, say, a hundred thousand years ago, I would expect some perceptible divergence from our present North-European human type; and there is none. We could not discover any difference between the Janos people and ourselves. Indeed, the medical examinations carried out on Earthfolk in their spaceships seem to have only one purpose: to reassure them, in scientific terms, that their type and ours have not significantly diverged during their long absence from Earth, and to confirm what they already felt emotionally and by their ordinary senses, that they and we are kith and kin – "*You are our people, because you are the same as us*".

Uxiaulia did, in fact, tell Frances that, medically speaking, they could find only one very small difference between their people and ours: the Janos people in the ships have a slightly higher average pulse rate than ourselves; but he said they think that, once they had lived on the ground for some time, even this would automatically adjust itself.

They seem, in fact, to set great store by this question of physical identity between them and us; as if it were a matter that they had worried over, and were happy to be reassured about.

Perhaps, one day soon, we shall be able to ask the Janos people about their history and origins; in what they have communicated to our English family in this one incident, the subject was not mentioned, except by inference: "*To us, you people are living history*".

They have made it clear that they know they are our kindred; but as to how and why they left Earth, nothing was said. I have wondered whether they themselves know the answers to these questions.

The 'why?' is almost more puzzling than the 'how?'. What driving motive lay behind that tremendous journey, so far into deep space?

Again, if they were involuntary passengers, it would not be their motive which sent them to Janos, but the motives, totally hidden from us, of those unknown beings who lifted them off from Earth, and transported them to a planet like enough to Earth – though it meant a considerable journey – to give them a fair chance of establishing themselves.

If their passage to Janos was involuntary, was it part of some grand cosmic design? Was Janos in need of intelligent creatures to colonise and develop it? Who made the decisions? Whoever made the decision to send humans to Janos, made a mistake, or had insufficient information; for it was inexorably written by astronomical forces into the future of Janos, that the planet would eventually become uninhabitable, and that the rain of rocks would drive away the survivors to seek a planetary home elsewhere.

Again, if they were voluntary passengers, the same

questions arise; except that they would have to be given a compelling reason why they should undertake such a journey.

Was this a rescue operation? Were they lifted off to escape being frozen to death in an Ice Age?

On the theory that they were passengers, whether voluntary or involuntary, the higher civilisation which took them to Janos may well have given them a helping hand, to enable them to establish themselves; and they may have been taught science and other useful knowledge by their hypothetical mentors. If so, it would go some way to explaining why they are now so far ahead of us, in spite of having lived through something like ten thousand years less than we have; the several thousand years of external elapsed time, occupied by a journey in which they lived subjectively through two years only, was so much time lost out of their history, and they did it twice – there and back. On their time account, one would expect them to be behind us stay-at-homes, whereas it is we who are behind them.

What of the remaining possibility – that in the Earth's remote past, a group of Europeans of Nordic type somehow had a compelling motive to emigrate to a distant world, and had the technical means to do it unaided: can we reject this as too unlikely?

Certainly we cannot rule it out altogether; though it would require the disappearance without trace of a scientific civilisation far more developed than our own. As I have suggested, a late Ice Age could have obliterated such a civilisation if it were localised within northern Europe; glacier ice in large enough quantities will grind any artifact to powder, and the subsequent floods would wash out to sea anything that remained. A threat from advancing ice could also provide a motive for escape; though to leave the planet altogether seems a drastic remedy, seeing that mankind did, after all, survive a whole series of glaciations, right through the Pleistocene; though each exposure to Arctic conditions would have represented a serious cultural setback.

It would surely have been easier to move south into more temperate regions to escape the ice, as our own ancestors did, returning north when the ice retreated, in common with all animal and plant life. Given a really high technology, they could no doubt have hung on through the Ice Age by going under the ice, and living on the produce of the sea under the ice floes.

The interplanetary move seems unnecessary merely to escape an Ice Age; if they went on their own to escape a terrestrial disaster, it must have been because they expected something to happen which would render the whole Earth permanently uninhabitable. If they had spaceships, and the problem were a temporary one, they could have lived in space for a time, as the Janos people are doing now.

There is another difficulty: if humans left the Earth on their own technological resources, for whatever motive, they still could not have known that the planet Janos existed. Why undertake such an immensely long journey, as a shot in the dark? If they wanted to settle and colonise, they must have gone past scores of suitable planets on the way. No doubt Janos offered ideal conditions for settlement: but how could they know it was there? However advanced their astronomy, they could not have known the living conditions on a planet thousands of light years away.

They would be unlikely to find Janos by chance; and it would take too long. A journey of such a length is, for reasons which the reader will perceive, scarcely feasible except on the understanding that it is done in one unbroken flight at maximum speed, very close to the speed of light. There would be no opportunity for exploration and stopping by the way: any such diversion would lengthen the time required to an impossible degree; for every stop to look at a planet would involve deceleration to stop, and then acceleration again to close to the speed of light. Such a stop-and-go progress would take tens of thousands of years – unless they possessed at that time the mastery over inertial mass which the modern flying saucers do appear to have. Even so it would still take too long.

No question of it: they went straight to Janos, and somebody knew where they were going, and had a reason for going there. Any other journey, which other groups may have made to other, equally distant planets, would have to be on the same top-speed, non-stop basis, with a destination known in advance.

It does begin to look very much, to me, as if the journey from Earth to Janos was organised by someone else.

That 'someone else' had a motive for intervening in the affairs of terrestrial men and dogs goes without saying: but what that motive was, is totally hidden from us.

Whatever the motive, whatever the plan, it seems likely that the same net, so to speak, picked up at least three groups of creatures from the Earth, probably on the same occasion, something between ten thousand and a hundred thousand years ago. These were: first, people of European race, members of the present-day human species *Homo sapiens*; second, people of another species of *Homo*, closely related to, but distinct from *Homo sapiens*, and less advanced culturally; thirdly, a kind of wolf with thick, shaggy dark hair, possibly attached to the humans in a 'hunting-dog' capacity.

The presence of dogs in the Janos story weighs heavily on the side of a 'passenger' theory of their departure. A group of human astronauts, planning a space exploration journey in their own ships, would hardly be likely to take with them a breeding population of big, powerfully-built dogs little removed from the wolf; it would create too many problems, and might well be risky.

But a gigantic spaceship of a vastly superior culture could easily accommodate men and dogs, and keep both under control. To take the dogs at all suggests that the men were, at that time, primitive hunters who would not be parted from their dogs, not being able to imagine a situation in which the dogs were not indispensable; and their unknown captors indulged them in this, to the extent of taking the dogs along.

If – and I am still, personally, uncertain which of the two alternative theories is the correct one, though the weight of probability lies with the 'passenger' theory – if the ancestors of the Janos people were lifted off from the Earth's surface by another and much higher culture, they learnt much from it. They were left, also, with a clear tradition of Earth as a planet where they would find people of their own kind, and with a knowledge of how to return to the Earth if they should need to.

Whether their ancestors were capable of interstellar space-flight or not, when they left the Earth, it is certain that the

Janos people are capable of it today, with all the complexities of galactic navigation. One imagines that the big horseshoe-shaped desk which John saw, of which he was told: *"This is for navigation"*, in spaceflight showed star-charts upon the screens in the left-hand, smooth surface, where he saw a film. Incidentally, a number of observations have confirmed that the Janos people in the spaceship are generally right-handed, as are the majority of ourselves; their equipment assumes right-handedness, and people touching buttons on a switch-board – a thing they seem to do constantly – do so with the fingers of the right hand.

* * *

As the great migration fleet began to approach the neighbourhood of the Solar System, which on their own statement they had never visited in modern times, the responsible officers must have felt the need for information about Earth more up-to-date than that in their history books.

There are quite a number of late nineteenth century reports of 'airships' not of any known terrestrial origin. I am beginning to wonder whether – while I think we must dismiss all earlier sightings as 'non-Janos' – we were starting to receive visits from advance scout ships, travelling ahead of the fleet. I have already indicated that the 'small moons' of the 1888 diary, crewed by men with shiny clothing and with their hair cut very short, are very likely to have been from Janos. It is all in character.

It is in character, also, that they should take the trouble to repatriate the near-man colony, the hairy folk who went to Janos with them, and who would have died if they were not brought back to Earth. Is this what lies behind the qualifying word 'most' in Uxiaulia's statement: *"Most of our people could infiltrate into your population, undetected"*?

Such advance scout ships would have left Janos up to a century earlier than the main migration fleet. They might well have begun the business of sasquatch-repatriation, as well as initiating the intensive survey of modern terrestrial mankind that has kept the flying saucers busy right up to the present day.

It has been noticed that all the localities in which sasquatches have appeared – often loosely associated with UFO occurrences – have been such as would give them the most favourable chance of establishing themselves: remote mountain forest country with plenty of caves, most usually; though one group of reports comes from the Everglades in Florida. It is curious that the sites are, as far as I know, all in North or South America. It has been suggested that the Himalayan yeti is, so to speak, a sasquatch; but there is no clear link with flying saucers, and it is more likely that the yeti has been here all along.

For advance scout ships to be useful to the fleet command, to help in planning, it would be necessary to send information back to the approaching fleet. This could be done, no doubt, with a powerful tight-beam radio transmitter, which can be received over immense distances. Alternatively, some of the earlier-arriving advance scout ships might have returned to join the fleet, bringing samples and photographs, and the advantage of direct consultation without a time-lag.

The earliest arrivals, too, might not appreciate having to wait around for a century or so for the fleet to arrive. They could travel in their ships almost as quickly as a radio signal could go, though time would be lost in acceleration and deceleration, and in the complicated U-turn to match flight-path with the advancing fleet; but no doubt it would be worth while. Later arrivals among the scouts would remain here and wait for the fleet, gathering information and sending back reports by radio. I suspect that the 'foo-fighters' of the second world war, familiar to air force pilots both in Europe and in the Fast East, were observation drones controlled by ships in this category.

I am dipping rather deeply into conjecture; but it all seems to me reasonable conjecture, based on what we know.

* * *

I have indicated in the Preface that this book makes no claim to solve all the problems of the UFO phenomenon; it is confined to giving an account of the story of the Janos people as it was told to an English family, who were for nearly an hour guests aboard one of their spaceships. I have been careful to add to this factual account only such information as seems to me to follow logically from it, bearing in mind certain background knowledge.

I have mentioned the sasquatch business at some length because, if Natasha's report is correct – and I see no reason to doubt it – it is part of the Janos story and throws some light upon it.

On the other hand, sasquatches apart, descriptions of UFO occupants, and of beings associated with UFO landings, have varied over a fairly wide range of types, far different from the strictly Nordic European type of Janos, as shown to us in this incident. There have been a great many reports of small people, between three and five feet high, with a great variety of clothing, equipment and spacecraft, though the craft vary much less than the people, and give the impression of being all derived from one technological system.

There have also been reports of 'giants', though I am inclined to discount many of these. When I find a six-foot man, or even a seven-foot man, described as a 'giant', I give up; there are plenty of seven-foot men on Earth today, particularly if one allows for a little overstatement. Where we have sincere reports of 'floating' figures twenty feet high, of terrifying aspect, I suspect we are dealing with projected images, intended to frighten people away from something which the real people, of normal size, did not wish to be observed.

The Janos story, as it has been told to us, has nothing to say concerning the small people, and certainly nothing about the more extreme oddities of the goblin or leprechaun type. There are too many 'little people' reports to discount; there must somewhere be a basis in fact: but from this present incident I cannot offer an explanation.

Only one rather thin line of evidence suggests a link with Janos. Of the four 'alien' pictures shown to Natasha by Akilias, one, the creature called Phusantheas and described by Akilias as 'friendly', is drawn by Natasha as a humanoid – a distinctly goblin-like creature with long ears, and coloured green all over. She shows them in her sketch as a group among houses, surrounded by some very odd-looking trees; there is a spaceship in the picture. The other three aliens – Vonason, Saunus and Faun – are less clearly drawn, but appear to be non-human.

In so far as reports of 'humanoids' – man-like creatures, not actually human in the narrower sense – are genuine, I suspect that their ancestors all came from the Earth, and that the original reality behind the reports is drawn from various members of the hominid Primates; that is, side-branches of the human family and distantly related to ourselves.

The sasquatch story may be only one of many similar stories, of semi-human and near-human creatures that evolved upon the Earth, way back in Tertiary times. Most of them have died out, and are known only from fossil remains; a few may still survive in wild and hidden places where they are seldom encountered. It is possible that a few may have been taken to other planets, and have returned; but whether in Janos ships, or otherwise, I do not know. Only the 1888 diary clearly indicates a Janos spaceship bringing sasquatches 'from the stars'. Whether they travelled, or whether they stayed on Earth, they are all our distant cousins, and originated in this planet.

Terrestrial mankind has yet to meet, face to face, an intelligent being – or a being of any kind – which originated in another world. When that day comes, we shall have a surprise; I hope it will be a pleasant one.

CHAPTER FOURTEEN

Homecoming

So THE GREAT migration fleet is here – somewhere in the Solar System. It is understandable that they should not wish to be more specific than that: the Janos people, however friendly they may have been to our visiting family, are not naïve; and indeed they have shown acute perception of the political atmosphere of Earth – *"Some people want to use our knowledge to control other people"*.

Their presence in the Solar System, and their expressed desire to settle permanently on the Earth, creates a political hot potato of the first magnitude – if you permit a vegetable to have an astronomical attribute. For one thing, there is going to be the very awkward question: where are they going to live? This one is so tricky that perhaps we should defer it to later in the chapter; but as a problem it won't go away.

"We come in peace, if all agree": this characteristic Janos phrase reflects accurately their own political atmosphere – they are essentially democratic, much more so than any Earth nation is in practice, whatever the political theory. The stupendous ring-shaped 'flagship' (my term, not theirs) is *"where all reports go, and where the big meetings are held"*, as well as being a sort of Grand Central Station for the fleet: for it is ringed with huge entry ports, so that ships can enter and leave, many coming and going at the same time if need be. It is quite certainly the largest artifact in the Solar System.

"We do not make the decisions; we have to report everything back. All must be in agreement. There will have to be a great deal of discussion before anything is decided."

One has the impression that the Janos people themselves do a fair amount of talking; and no doubt, once we get down to detailed negotiation, they will hold their own. There must be many who, like Uxiaulia, speak fluent and idiomatic English; and incidents reported from other countries have indicated that some of the spaceship people can speak Spanish well, for example.

"*Our ship is one of those which have been chosen to make the first contacts.*" And who have been among their first contacts?

Answer: John, Frances, Gloria and two little girls – Natasha and Tanya. Ordinary folk: as one who knows them well, I may say very good examples of ordinary folk; but ordinary none the less. That is what the Janos people seem to be interested in. John put it with unnecessary modesty: "Why me?"

One hears frequently the question: "Why don't they talk to Chairman Hua, or Mr Brezhnev, or Jimmy Carter, or somebody important?"

The answer seems to be, as far as we can judge, that leaders and governments frankly do not interest them; and who is to say they are wrong?

The Janos people have a profound sense – which we would all do well to copy – of the importance of ordinary folk, and the right of ordinary folk to have a direct voice in the making of policy decisions.

How many people take part in the 'big meetings' in the flagship, I do not know; but their mastery of electronic communication, and their constant use of visual aids – this has also been reported from several other incidents – could make possible the direct personal participation of very large numbers of people.

Talking of numbers, this is a matter which is going to raise political eyebrows. With their fondness for rather imprecise general expressions, they have told us no more, concerning their present total population, than that "*there are enough of us to fill one of your large cities*". I am taking this, thinking in terms of New York, Moscow, Tokyo, London, as being of the order of ten million.

There has been, of late, so much fuss about finding homes for a few thousand refugees, that ten million is not going to be easy. Uxiaulia told Frances that: "*it would be possible for most of our people, in very small groups, to infiltrate into your population, undetected"*; and Frances had a strong impression that they spoke from experience.

But he also made it plain that such is not their desire; they wish to come openly and by general agreement. They want to have *"a place which is our own, where we can all be together, and be independent"*. Ideally, they would very much like to have an island of their own; it would need to be quite a big island.

There are many questions of a more or less technical nature which will have to be gone into, once we reach the negotiating stage. But we are not there yet: the first issue is the issue of goodwill and mutual confidence; and this is not going to be an easy one.

We of this planet have been brainwashed for centuries, by writers long before H. G. Wells wrote *The War of the Worlds*, and more lately by film and television, into assuming – as in any case we all too readily do – that people from other worlds are necessarily sinister and hostile. Though it would be difficult to imagine anything more sinister and hostile than some of our home-grown villains.

Only an hour before writing these words, I was watching a television series, clearly of American provenance, called 'Project UFO'. It was well produced, and made use of some impressive techniques: but its purpose was a little obscure, since it seemed to be trying to suggest (a) that UFOs are hostile and will attack people and harm them, and (b) that UFOs do not really exist anyway, but are all in the imagination. I found it difficult to reconcile (a) and (b).

Then, still thinking of mutual confidence, try to look at it from the point of view of the people in the spaceships; and don't forget that, so far as the Janos people are concerned, though they do come from another world, they started here, and think as we do. It is not surprising to learn, though it makes one feel sad and ashamed, that they are all the time terrified of being captured by the savages on the ground.

No, 'savages' is not their term; it is mine. Race-hatred, religious murder, pogrom, the gas chambers, Belsen yesterday, Indo-China today – need I go on?

Mercifully, the Janos people are intelligent enough not to

generalise, and are capable of realising that ordinary folk like John and Frances – ordinary folk in any country – are not the makers of horror and terror. These things are done – bargaining with human lives at gunpoint, capturing women and children to extort ransom-money on pain of death, blowing up innocent people who have done no harm, raping children, selling narcotics to teenagers – by monsters from another world, is it?

Which are the humans? These savages lacking compassion or a sense of humour; or the people in the flying saucers, who can say in all sincerity:

"Please, you must not be afraid at all. We mean you no harm whatsoever."

The Janos people are human; this we have established, in this enquiry, beyond any reasonable doubt. In many ways they are more human than some of those who never left Earth.

But had they turned out to be aliens with green skins, big round froggy eyes and slit mouths, like some of those who have been described in UFO reports, would that have made them unacceptable? Should we not judge people by character, kindness, intelligence, humour, rather than by an anatomical description? I would rather deal with a 'monster' with a kindly twinkle in his eye than with a heartless, humourless man.

Humanity is a quality of mind. 'Human' and 'humane' are closely related in meaning. It was the Scottish poet Robert Burns who spoke bitterly of "man's inhumanity to man"; we have plenty of examples around today.

Perhaps the Janos people, human with a different history, can help us to remember what 'human' really means. An example and a strong lead, backed by the authority of great knowledge, could really start something in this tormented planet – a movement of ordinary common folk which would put an end to cruelty and inhumanity, and would assert the right of ordinary private people to decide how our world should be ordered. A movement which would keep political, military and economic power in check, and uphold the freedom and dignity of the individual. A movement which would do away with labels and categories, like 'Jews', 'Arabs', 'blacks', 'whites', 'reds', 'women', 'bosses', 'workers' – and leave us as just plain John Smith and Mary Jones. No titles, no labels; just people. No masses, no categories; just individuals.

The only terrestrial organisation I know of, though I am not a member of it, which really thinks like this is the Society of Friends, commonly known as Quakers. It is surely not an accident that this Society is known throughout the world for its practical humanity. I have a feeling that the Janos people and the Quakers would find much in common; though the Janos people who talked to John and Frances said not one word about religion. They have not come with a message or a mission; only they need a home, and have come to us, their kinsfolk, in their need.

Race prejudice is unhappily with us, and cannot be ignored. I have an uneasy feeling that there are going to be large numbers of Earth people who, perhaps subconsciously, will start off on the wrong foot in their attitude to the Janos people; because the Janos people, as far as this and many other reported incidents tell us, are 'white' people; they are, without any doubt, Europeans. They resemble most closely the Scandinavian nations; though their language, from the little we know of it, is more like Greek.

The two do not necessarily exclude each other: we are talking of an age, when the ancestors of the Janos people lived on the Earth before, of at least ten thousand years ago; I think we have in the Janos people an ancestral form, from which is derived the tall, blond, slim type, with wavy flaxen hair and ice-cool blue eyes, which has always been accepted in European folklore as the ideal type – the fairytale type.

As to the possible Hellenic link (the very word Janos has a distinctively Greek ring to it; and Saton is actually a Greek word), let us remember that, while present-day Greeks conform, in general, to the Eastern Mediterranean type of European, their classical art and ancient mythology refer back to a blue-eyed fair type as an ideal – a type identical with the Janos people of today.

It is perhaps not just a coincidence that 'democracy' – a word which essentially describes the Janos political point of view – is a Greek word and an Athenian concept; though we should remember that the world's first parliament was in Iceland, and spoke Old Norse.

Scholars know that peripheral cultures, by which they mean people who have moved away from the ancestral centre, have a tendency to be conservative, retaining some older forms of the central culture, long after the centre itself has moved on to new forms. The slight archaism of the French language of Quebec and the Seychelles is a case in point.

It will not greatly surprise me if the Janos language, when we come to know more of it, will turn out to be allied to an archaic proto-Hellenic. If I had to suggest a likely place of origin for the Janos people, I would be inclined to put it in Eastern Europe, perhaps in what is now Bulgaria or Yugoslavia. From there, I would expect to find that some of them moved north-west into the Baltic and Norse countries as the ice retreated, while others moved south into Greece. And one group, by what means we do not know, left the Earth altogether for a distant world, and are now seeking to return to their world of origin – their birth-planet.

Where are we going to put them? Or rather, what place should we offer? – for the Janos people have been doing their own surveying, and may have something to propose.

As soon as I realised that a home was needed for something like ten million people, with a strong desire for an island of their own, I went straight to a good atlas. I did not need to open it to know that the problem was going to be difficult: that there is a shortage, particularly, of suitable large islands which are not already overcrowded.

It must have been something of a shock to the Janos people, accustomed to having plenty of room on their lost planet, to discover that the human population of Earth had grown to around four thousand million, and is rising steeply. The Earth takes in more extra people every month than the total number of the Janos people.

The land surface of the Earth, with 25 persons to the square kilometre or 64 to the square mile, is however not overcrowded as a whole; the world divides rather sharply into high-density areas and low-density areas. Broadly speaking,

Europe and Asia are high-density, though there are large areas of low density in the North; while Africa and (more surprisingly) America, North plus South, are low-density, well below average.

But I am not ready yet to make named suggestions. Let us look first at the advantages of providing 'a home and friendship', as they ask us to do, for the Janos people: I mean the advantages to us, to the present population of Earth.

I would like to think that the human race would be capable of moving over and making room for the extra Janos population, purely from humanitarian motives: because they have had much trouble, and need a home – need it rather desperately: *"we cannot float around in space for ever"*.

Frankly I have not that much confidence in our humanity. We would need to be paid.

Fortunately for the Janos refugees, they are able and ready to pay for the land we give them, many times over in value.

Through John and Frances, they have proposed a trade. Knowledge for land. Science for territory. Technology for acceptance. They would hope also for our friendship; but that cannot be purchased.

With Janos science, technology and experience, grafted on to Earth's enormous resources, all space would be open to us. The Galaxy would become the new frontier.

With the Janos ability to control gravitation, to make things weightless, Earth engineering would be transformed. Cars that float above the ground; 'anti-grav' devices for lifting impossibly heavy loads – these are only two of the immediate applications that we can vouch for in this one incident.

Their medical science, immensely superior to ours, would be open to us. The medical rooms in the spaceship, which John and Frances describe, contain an intricate complex of electronic instrumentation; though it is clear that they used at least one drug administered by mouth in this incident, to facilitate hypnotic control.

Where the Janos people are a very long way ahead of us is in mental science. In this connection, both John and Frances were greatly impressed by the relaxed, unstressed personalities of the people in the spaceship – 'a built-in tranquillity' was one phrase used. John said there was no hostility. The Janos people's own mental health is clearly in better shape than ours; all this we can benefit from.

They could help us to solve our energy problems: though I think we all know, deep down, that the remedy is in our own hands – we use far more than we need.

"*You seem to enjoy doing things the hard way here*": Janos technology could give us what they already had on the planet – the abolition of drudgery and hard labour. "*You should let your machines do more for you.*" It was in the context of domestic technology that Uxiaulia told Frances: "*We are far more automated; we are two hundred years ahead of you*".

More and better science: that is what we need; and the Janos people are ready to teach us their science – centuries ahead of ours – in return for a place to live, and a home in this world, which was once their world.

One thing we should not expect from the Janos people. It has become widely known – though the authorities concerned have tried to keep it hidden – that both Soviet and American military leaders hope to discover the secrets of the spaceships, to use them against each other. The Janos people themselves have told us: "*We can fight no more wars*". It is unlikely that they would be willing to sell military secrets to East or West.

But if the Earth–Janos combination is attacked – as the Janos people have warned us could happen – by ships from hostile worlds, their previous experience of such attacks could help us. However, the threat is not thought to be imminent: "*too far to do much harm at the moment*", they tell us.

"*We could learn from each other.*" Unnecessary modesty on their part, perhaps; but no doubt it is true. Earth may have a lot to teach Janos, possibly in the field of the arts, entertainment, music, dance, the pleasant things of life. We are very good at these things, even if our technology is mediaeval.

The mutual stimulus of new ideas, new points of view, comparing notes, pooling experience, could lead to an unprecedented flow of creative energy in both them and us.

And, of course, most important of all, sooner or later we shall have to begin to drop this 'us and them' business. We are all humans, more or less.

"*Together we can conquer all space*": yes, in time, in the sense that Frances felt, of peaceful expansion into the Galaxy; but first we must conquer our fears of each other.

Having overcome our xenophobia, our distrust and fear of strangers, on the interplanetary scale, we might bring the cure a little nearer home, and agree that Americans and Russians and Chinese need no longer fear one another. We all want to do it, all of us ordinary people – to make a world in which the fear of war has vanished. It only needs a little confidence, and the breaking of some bad old fixed habits of thought, which perpetuate distrust.

Something like half the world's resources go on so-called 'defence'. We do not really need the help of the flying saucers to stop this appalling waste, which is the main cause of our backwardness and poverty; but if Janos can help us by example, and by bringing into the world a new respect for humanity as a quality of mind, then we should be grateful.

What the Janos people are proposing, in effect, is a treaty of friendship and cooperation. This will have to be formalised, however it is negotiated, by the United Nations Assembly, and signed for Earth by the Secretary-General. No other organisation has the authority to speak for our planet. It will be the first time a decision will have been made by the people of Earth as a whole; and an attempt should be made to associate with the decision any nations which are not members of the UN.

It is therefore necessary that the individual inhabitants of Earth should all, as far as possible, be fully informed as to the nature of the issues involved. The publication of this book makes a modest beginning of this process of world-wide consultation. The question of the re-admission of the Janos people to Earth must not be decided by 'important people' behind closed doors; it is a matter which concerns us all.

Particularly, it must not become a political football between East and West. In fact, the matter of the Janos people could well be one in which East and West find a common approach; I trust that it will not be a negative one (there is a real danger of this).

Once the Janos people are settled on Earth in a country of

their own, they will at once qualify as a sovereign nation-state, for diplomatic and other international purposes. Janos should, as a matter of course, be invited to become a member of the United Nations; and I would personally think that their extremely advanced science and technology, and by implication their enormous military potential, would make Janos a suitable candidate for a permanent seat on the Security Council. We may as well recognise at the outset that, despite their comparatively small numbers, Janos will soon acquire 'great-power' status, and will form a new and influential element in world affairs.

I would suggest that it may be convenient, in the earlier stages of the settling-in of Janos as an Earth nation, for one established and trusted nation to assume the quite temporary role of Protecting Power; though full independence should not be delayed longer than is absolutely necessary. It is important that the nation chosen for this role should not be one of the present Great Powers. As soon as the advantages of the position are realised, there will be a rush of candidates for the post: the United Nations should not be hasty, or too conventional, in making the appointment.

The choice of Protecting Power will necessarily depend in part on geographical considerations; and this brings up – we can delay no longer biting into the hot potato – the question of Where?

Long hours of study of a good atlas have left me with little more than a headache for my trouble.

I have taken as a basis of calculation the somewhat arbitrary figure of ten million, as an order of magnitude for the present population of the Janos people in the ships. It is an interpretation of Uxiaulia's phrase: *"There are enough of us to fill one of your large cities"*.

Frances has a feeling that ten million may be an over-estimate; if so, a lower figure would make the problem of the re-settlement of Janos on the Earth that much easier to solve – though it will still not be easy.

On the other hand, we do know that Janos suffered a substantial loss of population in the disaster which drove them out of their planetary home; and perhaps we should make some allowance for recovery. We do not know whether the

Janos people are accustomed to regulate their population numbers by policy.

If one takes ten million, this is not a large population for an Earth nation, which is what Janos will become. It is comparable with the present populations of Australia, Hungary, Portugal, Sri Lanka, Malaysia, Kenya or Chile. It is less than a fifth of the population of the United Kingdom.

Now how much space will the new Janos nation need? Janos will clearly be an industrialised nation; but that does not mean that they should be expected to be overcrowded to the extent of Japan, with 700, or England with 900 people to the square mile.

The overall density of Europe, excluding the European portion of the USSR for which separate statistics are not available, is about 250 persons to the square mile. If we think of Janos as a nation of Europeans, at this density they would occupy 40,000 square miles or about 100,000 square kilometres. This is an area comparable to those occupied by Bulgaria, Hungary, Portugal, Liberia, Malawi, Jordan, Cuba or Guatemala. It is less than half the area of the United Kingdom. Even so, it is not going to be easy to find a tract of habitable land of that size going spare; especially there is a shortage of suitable islands as large as this.

The Janos people themselves, through Uxiaulia and Frances, have expressed a strong desire to have an island of their own, if at all possible. This is clearly a good idea from every point of view, but not an easy one to put into practice.

Islands of 40,000 square miles or more do, of course, exist in numbers; the trouble is that, with the exception of Antarctica, they are all inhabited.

The total area of New Zealand, for example, is about 100,000 square miles, carrying a population of about two and a half million. It is divided into two large islands, either of which would be sufficient for the Janos people at a European density. But New Zealanders would be unlikely to welcome the idea of giving up half their territory to newcomers, however civilised and congenial they may be; and this, I am afraid, is going to be typical of the responses of people who have islands big enough for Janos, which they could manage without.

I give New Zealand only as an example – though, objectively speaking, a good and favourable one – of a part of the Earth where, given goodwill and motivation, room could be found for Janos without serious loss or hardship to the displaced population. It could not be done all at once; and there would have to be substantial subsidies from world finance.

But there are other possible examples. If the Janos people do not insist on an island (though I can see their point of view – the island idea is a very sound one), there is plenty of room in the interior of South America, and also in Canada if they can adapt to the severe winter cold.

Australia, likewise, offers possibilities; the climate of the arid interior could be changed by a large-scale application of high technology – though I would like to see some thought given to the idea of a possible homeland for Janos in the North of Australia, having due regard for the welfare of the aboriginal population.

Africa would not be a good choice; because, although there is room to spare, it would create needless political frictions to try to introduce a large additional population of white people.

There may be possibilities in South-East Asia; but because I do not know the area, I leave it to others to offer suggestions.

There would, I think, be objections to a territory being offered to Janos which was directly controlled by the United States, the Soviet Union, or China; Janos as a nation must be given the chance of making its own choice of alignment, or of remaining non-aligned. The more independent Janos can remain, the more valuable its presence will be in the Earth.

There remains the continent of Europe, the continent from which the ancestors of the Janos people originated, long ago. One could look at it this way: where, from an ethnic and cultural point of view, would the Janos people feel most at home?

They are more like the Scandinavians than any other terrestrial group, not only in physical appearance, but temperamentally. One could think of the Janos people as 'space-Vikings' – wanderers on the galactic ocean, now returning home to the Northlands of their forefathers and foremothers.

Scandinavia has great numbers of islands; but only one of them is of any considerable size – Iceland, whose language and culture is perhaps closest to the Norse of antiquity.

What is perhaps more to the point is that Norway, Sweden and Finland, considered together, have a lot of spare space to the northward, and – what would appeal to the Janos people – any number of lakes and fjords. Like Canada, there would be the problem of learning to live through the winter cold.

I would like to see these three nations – Norway, Sweden and Finland – consider together whether they could find room between them for a new (and very old) Nordic people. Certainly there is room to spare to the northward; though due care would need to be taken to safeguard the welfare of the people of Lapland.

There would be a rightness in the Janos people living once more in Europe.

An open letter to the Janos People

JOHN, FRANCES, GLORIA and the two little girls Natasha and Tanya visited one of your ships, near Faringdon in Oxfordshire, England, on the date we call 19 June 1978. Anouxia talked to John, and Uxiaulia talked to Frances. Both were shown pictures of why you had to leave Janos.

They have told me (Frank, the writer of this book) about their visit, and about all you told them and showed them. John has described the big circular room where you make power, what we call the engine room, and he has told me about the pictures Anouxia showed him on the screen in the big curved table in the navigating room.

Frances has told me what Uxiaulia said, and about the pictures he showed her, of the rockfall, and of life in your planet before rockfall. Natasha has told me about the pictures Akilias showed her on a screen. Gloria does not remember much; perhaps you could now help her to remember, now that memory has come back for the others, as you said it would in time. Gloria now has another child, as perhaps you know.

They have not told other people; it was agreed that I should write in a book everything they told me about the Janos people, and this is the book. If there are mistakes in it, I am to blame.

We felt that it was important for ordinary Earth people, in every part of our planet, to hear the story of the Janos people, and to understand why you want to come here to live. We hope that this is in accordance with your wishes.

People are always afraid of what they do not understand; and when they are afraid they make wrong judgements. People on Earth call your ships UFOs or 'flying saucers', as I am sure you know; and they have many very curious ideas about them. Ordinary people nowadays have all heard about the flying saucers, and many have seen them; more than half the people think they are real ships, and come from another world.

Governments and military leaders, for reasons they have never explained to us, try to cover up the UFO question; and they tell us that the flying saucers, your spaceships, are not real things, but are only in the imagination. Most ordinary people do not accept this, and think the governments and military leaders are trying to hide something which they do not want ordinary people to know about. We, the ordinary people, do not like secrecy; and we think it is time the question of the flying saucers or spaceships was brought out into the open, and understood by everyone.

It is time that you showed yourselves more openly to the Earth people. I do not ask you to take risks; I understand you are anxious about your safety: but I am sure that arrangements could be made for you to speak through our television system, directly to all the people of this planet.

It is not going to be easy to get people to agree to give you a place to live.

This planet is very crowded, with four thousand million inhabitants; but there are places which are not crowded, where you could live. The difficulty is that every part of Earth – except Antarctica which is quite unsuitable for living in – has some people in it; and they are going to be unwilling to move from their homes. However, if they are given a sufficient incentive to move, they could in some cases be persuaded.

I have written about some possible places in my chapter 14: Homecoming. An island, as you yourselves have suggested, would be very good, because it would be a naturally defined area which would not change; land frontiers can be changed, either way. You will see that I have made one suggestion which is not an island – in the northlands of Norway, Sweden and Finland, countries which I myself know and love. If you already know English, it is easy to learn Swedish; but the Finnish language is difficult. However, many Finlanders can speak Swedish.

You will need quite a lot of space; all I know about your numbers comes from one thing said by Uxiaulia: *"There are enough of us to fill one of your large cities"*. I have taken this to mean about ten million; if your numbers are much less than this, it makes the problem of finding a place for you that much easier to solve.

How much space you need will depend partly on how you propose to live, and how much land you need for agriculture. It would be unwise to have too little space, because this might lead to disagreements later on, when your children grow up and may feel they have not enough room. Whatever is agreed should remain for a very long time, without causing further problems.

If you like the sea and ships, this will help understanding and friendship; because it is something that many Earth people will understand.

You will be able to help us a great deal with our science and technology, which is a long way behind yours. More knowledge must bring a better life for everyone. At the same time, we can perhaps help you in many ways; it may be that our creative arts are more highly developed than yours. The richness and variety of our cultural inheritance is quite remarkable.

In any case, we have numbers, and numbers mean strength; if Janos and Earth join forces, we shall be strong together, in case some of the unfriendly planets you speak of send their ships to attack us. But we hope not to have any war.

It would be good if your coming back to Earth helped the people of Earth to be more united among themselves, so that they did not waste half their wealth on buying weapons to fight each other with. The ordinary people do not want war; it gives too much power to those in authority.

Let us talk together. If you want to talk to me, John and Frances know where I live; and so does the publisher of this book.

A Note on Credence and Credibility

THERE WILL BE many who will be reluctant to accept this book as factual, for reasons which have more to do with their own personalities and psychological make-up than with the facts of the case.

We are all familiar with the story of the dear old lady who, seeing a giraffe for the first time at the zoo, said: "I don't believe it". Almost daily in some part of the world, people are seeing for the first time a flying saucer, spacecraft or some other UFO manifestation: many are convinced by seeing, sometimes too easily; but there are always some who, like the old lady, cannot accept what they see as reality – and of course, many 'sightings' are undoubtedly mistaken.

And yet, if you see an aeroplane, even one of an unfamiliar type, do you deny its reality? I suggest that it is not the strangeness of the flying saucers that make some unwilling to think of them as engineering hardware with real flesh-and-blood people in them: the UFOs are by now becoming so familiar through published pictures, that they are no longer so strange, and the term 'unidentified' in 'unidentified flying object' is fast losing its meaning. I have even heard of a UFO being "definitely identified as an unidentified flying object".

What puts the flying saucers in a class apart is the feeling that they are extra-terrestrial, from another world. And yet we ourselves, as soon as we set foot on the moon, are extra-terrestrial in a modest way. No one suggests that we should not 'believe' in Neil Armstrong or Buzz Aldrin, for all they travelled in a spaceship and wore helmeted suits, not unlike those sometimes described as worn by 'spacemen' from a flying saucer.

'Belief', in fact, is an inappropriate concept when we are talking of UFOs. Either the thing exists or it doesn't; no one's 'belief' affects it one way or the other. Belief is a state of mind, often with complicated psychological connotations. Often it is culturally conditioned, as with religious belief. It has no bearing on objective phenomena, which exist, or not, as the case may be, quite independently of what people believe or think about them.

Having spent very many hours listening to John and Frances, both under hypnotic regression and in normal conversation, I have not the least doubt that they experienced what they say they did. This is the 'face-value' hypothesis, which I have chosen to adopt, as the simplest, most straightforward, and most convincing.

If any reader chooses to prefer an alternative hypothesis, that is his or her privilege. But if he wishes to maintain his own credibility, he must be prepared to show us that his preferred hypothesis is more credible than the one I have chosen.

What are the possible alternative hypotheses?

First, he could prefer the hypothesis that the witnesses invented it all, for some reason known only to themselves, for they have not sought notoriety or reward. While I have learnt to have a great respect for them as personalities, I have to say that they simply did not know enough. They were relating matters, the significance of which often escaped them for lack of background information, but which made immediate sense to me. To take a single example: the thirteen words of the Janos language which they repeated to me hang together naturally in a linguistic family relationship; they could not have known this.

Again, the invention of the falling moon theme would require a considerable amount of astronomical knowledge, which they simply did not possess; though they are capable of understanding it when it is explained to them. I would not be so sure that they are yet quite clear about the relativistic space–time physics on which the journey from Janos to Earth depended; but Frances faithfully repeated Uxiaulia's admirably clear explanation of it.

Second, the author could have invented it. I only wish I could; if I had that kind of creative imagination, I could earn a large fortune as a science fiction writer.

Third – a favourite one, this – it was all in the mind of the hypnotist. Not all of it: because some of the most important facts came out in normal recall, on occasions when the hypnotist was not present. It is true that Geoff M'Cartney has an interest in ufology, and has read a number of books about it; but this has chiefly helped us by giving him a personal motivation to be willing to take on the case, and to work very hard on it; it has not contributed factually. Indeed, on the rare occasions on which Geoff's personal background reading may be said to have intruded, it was unhelpful and confusing.

Are these alternative hypotheses credible?

There is a possible further alternative, which cannot be dismissed so lightly; though after careful thought, I have to reject it as far-fetched and unsupported by any evidence: and the witnesses who, after all, are in the best position to know, will have none of it. This is the alternative, which has provided matter for several books, which regards all UFO manifestations as a kind of confidence trick played, not by human hoaxers, but by some icily remote and vast intelligence, cosmic in scale, which has a twisted sense of humour.

The notion has had some airing in the literature, that the UFOs are not what they seem: that the people in them are clever disguises, made to seem acceptable to us, for something very much nastier and infinitely sinister. The trick, it seems, is so cunningly worked that people like Frances, who is an acute judge of personalities, are entirely taken in.

Frankly, I find this one the most way-out and insubstantial of the lot; its exponents have given us no solid grounds for believing it.

Lastly, there are the various intellectual dodges.

'Subjectivity' is the one which has the widest currency at the moment. But does it mean anything?

All experiences have a subjective element; otherwise we could not experience them. I have no direct perception of the objects before my eyes: what my conscious personality has to work on is at best a mental construct, often distorted by what I expect to see, made up from data supplied by my optic nerves from the retinae. My view of the world is unavoidably indirect, and my subjectivity plays an important part in it.

But it would be arrant nonsense to deduce from this that the world before my eyes exists only in my subjective mind.

When John tells me that a particular piece of metal in the spaceship is shaped thus, and has these dimensions, I prefer the common-sense view that the piece of metal is approximately as he describes it; I do not find it necessary to take refuge in an escape formula, such as: "Well, he had an experience, which is real to him".

Formulae like this – and there are others, such as the 'parallel universes' one – are bolt-holes for people to hide their heads in – people who cannot face the uncomfortable, challenging truth, that the world we live in is larger than we thought.

Books for Further Reading

The following books are listed here for the convenience of those who would like to extend their reading in ufology and allied fields.

I must, however, make it clear that the inclusion of a book in this list does not imply that I am prepared to endorse its reported data, or to support its author's conclusions.

It would be invidious and impertinent if I were to attempt any kind of evaluation; all I will say is that all these books are on my own shelves.

BAXTER, JOHN and THOMAS ATKINS: *The Fire Came By*. Macdonald and Jane's; Futura 1977
BERLITZ, CHARLES: *Without a Trace* (an account of the Bermuda Triangle). Souvenir Press; Panther 1978
BERLITZ, CHARLES: *Mysteries from Forgotten Worlds*. Souvenir Press; Corgi 1974
BLUM, RALPH and JUDY: *Beyond Earth*. Corgi 1978
BOURRET, JEAN-CLAUDE: *The Crack in the Universe*. Neville Spearman 1977
BOWEN, CHARLES (edited by): *Encounter Cases from Flying Saucer Review*. Signet 1977
BOWEN, CHARLES (edited by): *The Humanoids*. Neville Spearman; Futura 1974
CHAPMAN, ROBERT: *UFO – Flying Saucers over Britain?* Arthur Barker; Mayflower 1968
HOLROYD, STUART: *Briefing for the Landing on Planet Earth*. W. H. Allen; Corgi 1977
HOLZER, PROFESSOR HANS: *The Ufonauts*. Panther 1979
HYNEK, DR J. ALLEN: *The Hynek UFO Report*. Sphere 1978

KEEL, JOHN: *Strange Creatures from Time and Space.* Neville Spearman; Sphere 1976
KEYHOE, DONALD E.: *Flying Saucers from Outer Space.* Tandem 1970
KEYHOE, DONALD E.: *Aliens from Space.* Panther 1975
LEONARD, GEORGE H.: *Someone Else is on our Moon.* W. H. Allen; Sphere 1978
PAGET, PETER: *The Welsh Triangle.* Panther 1979
PUGH, RANDALL JONES and F. W. HOLIDAY: *The Dyfed Enigma.* Faber and Faber 1979
RANDLES, JENNY and PETER WARRINGTON: *UFOs – a British Viewpoint.* Robert Hale 1979
SACHS, MARGARET with ERNEST JAHN: *Celestial Passengers, UFOs and Space Travel.* Penguin 1977
SHUTTLEWOOD, ARTHUR: *The Flying Saucerers.* Sphere 1976
SHUTTLEWOOD, ARTHUR: *UFO Magic in Motion.* Sphere 1979
SMITH, WARREN: *UFO Trek.* Sphere 1977
STANFORD, RAY: *Socorro Saucer.* Fontana/Collins 1978
STEIGER, BRAD: *Strangers from the Skies.* Tandem 1966
STEIGER, BRAD: *Mysteries of Time and Space.* Sphere 1977
STRINGFIELD, LEONARD H.: *Situation Red: The UFO Siege.* Sphere 1978
TEMPLE, ROBERT K. G.: *The Sirius Mystery.* Sidgwick and Jackson; Futura 1976
VALLEE, JACQUES: *UFOs: the Psychic Solution.* Panther 1977
WILSON, DON: *Our Mysterious Spaceship Moon.* Sphere 1976

Flying Saucer Review is a reputable international journal of ufology. It is published six times a year by FSR Publications Ltd, West Malling, Maidstone, Kent, England.

UNITED KINGDOM REPORTS OF CLOSE ENCOUNTERS and other major incidents involving UFOs, flying saucers or possible spaceships from other worlds, or of 'space' people encountered on the ground, may be sent in confidence to any of the following, who will arrange for investigation where it appears this may benefit our understanding of these problems:
NORTHERN ENGLAND Jenny Randles, 8 Whitethroat Walk, Birchwood, Warrington, Cheshire WA3 6PQ   Telephone 0925-824036
SOUTHERN ENGLAND Ken Phillips, 13 Falcon Avenue, Springfield, Milton Keynes MK6 3HG   Telephone 0908-678870
WALES Randall Jones Pugh, Parkland Place, St Bride's View, Roch, Haverfordwest, Dyfed   Telephone 043-784 246
SCOTLAND Alan and Trisha Price, Bluehouses, Hassington, Kelso, Borders   Telephone 05737-314

UNITED STATES OF AMERICA reports may be sent to:
  Center for UFO Studies, PO Box 1402, EVANSTON, Illinois 60201
  Telephone (312) 491-6666

Index

advance scout ships 172, 173
age of people 64
agreement, general 156, 176
agriculture 76, 148, 190
air force 16, 173
airlock 28, 38, 93, 101
Akilias 32, 130
aliens 45, 175
alternative hypotheses 191
altitude meter 101
amnesia 17, 21, 31, 108, 130, 132, 133, 136
amplifier, public address 93
animals 76, 148
Anouxia 52, 79, 90, 93, 97, 99, 107, 130, 143, 149
anti-grav 5, 117, 182
architrave, decorated 153
arts, creative 183, 190
asteroids 112
Australia 187

Babylonian tradition 159, 160
badges, insignia 59, 60, 88, 90, 149, 151, 153
balaclava helmet 29, 80, 149
balcony 27, 28, 52, 92
barbecue 68, 69
bargeboard, decorated 140
beam, downward 36, 50, 88, 124, 128, 131, 132, 153
beam, rotor 104
beam, scanning 24
big man 59, 77, 153
black in dress 70, 151
blood samples 90
boats on Janos 2, 70, 151
bodice and skirt 69, 151
bolt-heads 40, 88, 147
books 193
bruise-like marks 17
buildings on Janos 140

cabinets, white 96, 100, 103
café-like room 60, 65

captain of ship 79
capture-fear 178
car 14, 33, 39, 49, 68, 76, 93, 131
car park 132, 134
catwalks 104
CE4 16, 30
chain reaction 5, 123
chair, 'dentist's' 23, 25, 32, 53, 55, 56, 79, 82, 87, 88
chairs, ordinary 40, 60, 105
children as witnesses 32
clasp, silver 59, 149
climate of Janos 2, 114, 139
close encounters 16, 30
clothing, Janos 69, 70, 71, 73, 148
clothing, monk's habit 6, 121, 123, 150
clothing, silver 29, 36, 49, 57, 80, 148
coffin 6, 121
columns, pillars 51, 91, 96, 121
computer 79
contacts, first 67
corridors 59, 91
Cosentia 82, 153
cosmetics, make-up 75
cover story 12, 19, 20, 132, 134
craft, oval 118, 120, 123, 141
crazy-bear 161
credibility 191
crew of spaceship 49, 60, 96, 100, 129, 132, 148
crops, agriculture 76
curved floor 28, 52, 92
cylinder, luminous 78, 128
cylinders, rotor 96, 104
cylindrical spaceships 160

democracy 180
desk, navigation 107
diary 1888 161, 172
dimensions of spaceship 49, 51, 106
dog or wolf 3, 148, 163, 167, 171
dome lights, coloured 55
doors, doorways 25, 52, 73, 91, 126, 128, 129, 141
dreams, Frances 26

Index

dreams, John 22, 50
dreams, Natasha 29
drinks 31, 60, 130, 133
drugs, hypnotic 133, 182
dungarees 71, 150
dust, radioactive 5, 65, 121, 123

ears 57, 80, 149
Earth 2, 109, 139, 158, 163, 166, 171, 181, 184, 185, 190
electromagnetism 99
elevator, lift 39, 78, 91, 104, 107, 124
engine room 28, 51, 91, 96, 106
English speech 107, 154
entry into spaceship 26, 33, 37, 50
Europe, Europeans 156, 166, 169, 171, 180, 188
examination, purpose of 55
eyes 29, 36, 57, 157

face 54, 58
face-value hypothesis 191
Faringdon 8, 12, 13, 14, 19, 30, 132
fashion, women's 69, 151
Faun 44, 154, 175
films 3, 16, 64, 67, 109, 110, 143
finial, carved 73, 140
Finland 188, 190
flags 71, 151
flagship of fleet 2, 147, 155, 156
fleet, migration 2, 6, 66, 147, 158, 172, 176
float cars 76, 118, 141, 145, 182
floating sensation 13, 86
floating up or down 32, 50, 91, 104, 107, 124, 129, 131, 132
floor, curved 28, 52, 92
floral motif 70, 151
flowers 140
flying saucers 3, 12, 32, 34, 42, 49, 153, 159, 189
flying shields 158, 159
foliage 68, 114, 138, 140
food 66, 69, 76, 148
foo-fighters 173
Frances 8, 26, 48, 53, 63, 126, 189
fruits on Janos 68, 69, 140, 148
furniture 60, 78

gables of house 140
galaxy 137, 182
garden, Janos 73
giants 174

Gloria 8, 17, 32, 40, 126, 133, 189
gloves 58, 113, 149
governments 177, 189
gravitation 53, 77, 97, 109, 118, 124, 127, 138, 160, 165, 182
Greek affinities 154, 167, 180, 181

hair styles 57, 58, 71, 73, 76, 91, 149
handrails 28, 52, 92, 101
hatch doors 28, 38, 51
head-covering 70, 151
heart and pulse 55, 57
hedge in car park 132, 134
height of persons 80, 91, 156
hip-clasp 70, 151
history 77, 158, 168, 169
homes on Janos 1, 2, 73, 76, 140
Homo sapiens 171
house that wasn't 8, 30
humanity 179, 180, 184
humanoids 174, 175
hypnosis, regressive 22, 46, 65, 136, 155, 156
hypnotic control 83, 86, 133, 135

ice ages 169
identity with Earthfolk 52, 55, 167
illumination 69, 145
industry, Janos 145
inertial mass 127, 170
insignia 59, 60, 153
instrument panels 22, 54, 80, 107
interstellar spaceflight 159, 164, 166, 171
islands 178, 181, 186, 190
itching of skin 17

Janos people, the 1, 135, 136, 154, 155, 156, 158, 163, 166, 179, 180, 186, 187, 189
Janos, the planet 1, 63, 113, 114, 137, 168, 170
John 8, 18, 19, 21, 33, 49, 51, 79, 96, 107, 120, 126, 128, 130, 131, 132, 141, 143, 189

kissing 130

lakes 1, 68, 70, 188
lane, narrow interminable 13, 133, 134, 135
language, Janos 91, 154, 180, 181
laughter 1, 77, 97, 99

Index

leaves of plants 68, 114, 138, 140
ledge in airlock 37, 51
letter to Janos people 189
lever, long 56
lift, elevator 39, 78, 91, 104, 107, 124
lifting ship 77, 97
lighting, road 145
light over chair 56, 88
light, speed of 6, 159, 170
lights, ceiling 60
lights, ring of coloured 11, 35
light-years 6, 63, 137, 159, 170
little (green) men 45, 174
lower deck 93, 101

make-up, cosmetics 75, 157
M'Cartney, Geoffrey 46, 65, 192
meat 69, 76, 148
medical examinations 53, 79, 90, 182
medical personnel 23, 54, 57, 90, 91, 153
meetings 156, 176
mental science 135, 182
meteorites 65, 66
microphone 93
military leaders 67, 183, 189
mist on road 49, 50
motives for journey 168, 170

Natasha 8, 16, 18, 29, 31, 130, 133, 162, 174
navigation 107, 167, 172
New Zealand 186
Nordic, Scandinavian 156, 167, 169, 180, 187
Norway 188, 190
nuclear power stations 4, 122, 145, 155
null-gravity 77, 97, 99, 118, 124, 149, 160

ordinary people 67, 177, 179
oscilloscope 84, 85, 86
own-steam alternative 163, 169
Oxfam people 65, 121

passenger alternative 163, 166, 168, 171
pennant flags 71, 151
personality, Janos 156, 182
pets, animals as 76, 148
Phillips, Ken 30, 194
Phusantheas 44, 154, 174

physical type 57, 80, 91, 138, 156, 166
pillars, columns 51, 91, 96, 121
pipes, pipework 103, 104, 105
pivot, rotor 105
planets, inhabitants of 44
planets, pictures of 109
politics 156
population numbers, density 3, 67, 177, 181, 185, 190
portholes 59, 79
power, electrical 4, 77, 99, 100, 101, 145
power generators 77, 99
power stations, nuclear 4, 122, 145, 155
prehistoric near-humans 163, 167, 175
press-button devices 37, 40, 42, 54, 78, 96
protecting power 185
pulse rate, heartbeat 55, 57

Quakers 180

race 156, 180
radioactivity victims 6, 65, 121, 122, 123, 150
ramps 27, 28, 39, 52, 92, 126, 128, 129, 147
Randles, Jenny 30, 194
reassurance symbol 54
recap sessions 47, 86
relativity 63, 64, 157, 159, 192
rescue ships 4, 123
right-handedness 172
road, Janos 143
road, narrow interminable 13, 133, 134, 135
rockfall 2, 3, 4, 5, 65, 67, 115, 139, 168
rocks on ground 66, 115, 116, 117, 123
roofs of houses 73, 76, 140, 141, 143
room like café 60, 65
room, medical 53, 79
room, navigation 107
rotor 96, 97, 99, 103, 160
rotor beam 97, 104, 105
rotor deck 96, 103, 104, 105
rotor pivot 105

Sanskrit tradition 159, 161
Sarnia 113, 139
sasquatch 43, 161, 175

Index

Saton 1, 4, 113, 115, 139, 154, 164, 165, 180
Saunus 44, 154, 175
Scandinavian, Nordic 156, 167, 169, 180, 187
screen image quality 65, 68, 109, 122, 143
screens 38, 42, 64, 78, 101, 109, 110, 143
seasons 139
seat-belt 41
Serkilias 82, 88, 89, 90, 153, 157
settlement on Earth 176, 177, 178, 181, 184, 189
shields, flying 158, 159
ship 3, 19, 52, 107, 177, 190
shipyards, underground 3, 5, 120
shoes 58, 59, 73, 99, 149, 150
shops 143
Sirius 137, 160
skirt and bodice 69, 151
small people 174
solar system 167, 176
sounds of spaceship 12, 16, 97, 99
space 155, 182, 184
spaceship 1, 11, 16, 19, 49, 100, 107, 126, 127, 141, 147
speech, English 107, 108, 154, 177
speech, Janos 76, 91, 154
speech of welcome 14, 52
spiral escalator 48
stairs, steps 101
stars 110, 111
static electricity 68
stereo images 110
Sweden 188, 190
swim-trunks 69, 150

tables 60, 78
Tanya 8, 18, 31, 33, 40, 128, 130
telepathic communication 63, 65, 79, 155
time 15, 63, 64, 133, 135, 159
track-suit 71, 150
transformers 102, 147

transport, Janos 76, 141
trees on Janos 68, 140
Tunguska explosion 160
tunnels 3, 119, 120, 123

UFO 12, 14, 44, 67, 158, 173, 174, 178, 189, 191
UFO investigators' network 30
undercarriage, tripod 32, 43, 49
uniform of ship's crew 29, 36, 49, 57, 58, 59, 80, 148
United Nations 184, 185
upper room 78, 126, 127, 128
uranium 145, 164
Uxiaulia 63, 71, 76, 140, 149, 155, 164

vanished civilisation 164, 169
vegetation on Janos 68, 75, 114, 138, 140, 141
velocity of transit 6, 159
visual image transmission 155
voltage, high 99
voltmeter 100
Vonason 44, 154, 174
Vurna and children 67, 73, 140, 150, 166

wall, grey 104, 106
war 1, 67, 155, 183, 184, 190
water on Janos 2, 114, 115, 138
weightless state 77, 97, 124
welcome speech 14, 52
windows 34, 73, 132, 141
wolf-dog 3, 148, 163, 167, 171
women in spaceship 36, 41, 76, 80, 130, 156
women's clothing, Janos 69, 71, 73, 148
working day, Janos 147, 183
writing, Janos 89, 153

xenophobia 184

yo-yo movement 11, 99

zip fastener 59, 149